高等教育应用型人才“十三五”规划教材
高等教育应用型人才计算机类专业规划教材
本教材学习网站：www.lgsworks.com
江苏高校品牌专业建设工程资助项目

C/C++程序设计实用案例教程

丁　展　梁颖红　李广水　主　编

電子工業出版社
Publishing House of Electronics Industry
北京 • BEIJING

内 容 简 介

本书主要利用C/C++语言进行综合课程设计学习。书中案例涵盖了字符串处理、排序与查找、栈、链表与队列、树与图、递归与分治、集合与映射，以及Win32 GUI编程基础等知识。在本书的最后给出了四个综合课程设计的开发案例，每个案例都提供了完整的源代码，结合书与示例代码，读者可以很容易理解。

本书可以作为应用型本科院校程序设计课程的教材，也可作为软件开发人员的指导用书。

图书在版编目（CIP）数据

C/C++程序设计实用案例教程 / 丁展，梁颖红，李广水主编. —北京：电子工业出版社，2018.8
ISBN 978-7-121-34603-3

Ⅰ. ①C… Ⅱ. ①丁… ②梁… ③李… Ⅲ. ①C语言－程序设计－高等学校－教材 Ⅳ. ①TP312.8

中国版本图书馆CIP数据核字（2018）第142582号

策划编辑：李　静
责任编辑：朱怀永　　　　文字编辑：李　静
印　　刷：北京七彩京通数码快印有限公司
装　　订：北京七彩京通数码快印有限公司
出版发行：电子工业出版社
　　　　　北京市海淀区万寿路173信箱　邮编100036
开　　本：787×1092　1/16　印张：14.75　字数：377.6千字
版　　次：2018年8月第1版
印　　次：2018年8月第1次印刷
定　　价：42.00元

凡所购买电子工业出版社图书有缺损问题，请向购买书店调换。若书店售缺，请与本社发行部联系，联系及邮购电话：（010）88254888，88258888。

质量投诉请发邮件至zlts@phei.com.cn，盗版侵权举报请发邮件至dbqq@phei.com.cn。

本书咨询联系方式：（010）88254604，lijing@phei.com.cn。

前　言

随着信息技术的迅猛发展，程序设计在各个领域都起到举足轻重的作用。目前大部分高校的计算机、软件及相关专业都开设了各种程序设计和数据结构课程。然而掌握一门程序设计语言或者了解数据结构知识并不等同于“会编程”。在实际教学中存在大量的只会考试但不会写代码解决问题的学生，这是令人焦虑的现象。

虽然已出版的关于算法设计的书籍比较多，但大都介绍算法的理论基础。对于实际生活中的各种问题，需要灵活应用算法。本书精选了大量综合编程案例，基本涵盖了当前基础算法领域的热点问题。

本书的特色如下。

一、本书提供大量的综合编程实例，涉及各种基础算法的应用领域。

二、所有的综合编程实例都按照设计思路、相关概念、原理、具体实现进行介绍，学生能够很容易地掌握实例的算法设计思路。

三、本书介绍图形用户界面(GUI)编程基础，将四个综合编程实例的算法进行可视化，从而让读者理解并掌握算法可视化方法，进一步帮助学生理解与调试算法。

要运行本书中的绝大部分实例，只需要安装任何 C/C++开发工具，如 CodeBlocks、Visual Studio 2015 社区版、Dev C++等。如果要运行四个算法的可视化实例，则需要安装 CodeBlocks 或 Visual Studio 2015 社区版。以上工具均可以免费下载。如果计算机运行的还是 VC ++6.0，我们强烈建议卸载。因为它的编译器版本太低，不支持 C/C++新标准，并且对模板的支持不够。

本书共有 8 章，内容如下。

第 1 章字符串。主要介绍字符串复制、连接、与数的转换、查找、删除，以及字典比较等各种字符串操作。其后，给出若干字符串处理常见问题的解决方法。本章最后给出综合编程实例：公民数据模拟。

第 2 章排序和查找。主要介绍桶排序、qsort 排序、STL 中 std::sort 排序、std::greater 用法、二分查找算法、std::find 查找用法，并给出两个综合实例：员工 KPI 排名与 MOOC 期终成绩。

第 3 章栈、链表与队列。主要介绍 STL 中 std::stack、std::list、std::queue 与 std::deque 的用法。对以上数据结构分别给出了若干综合编程实例，如关于栈的火车调度问题、链表的约瑟夫环问题、队列的卡片游戏等。

第 4 章树与图。树的内容主要包括完全二叉树定义、二叉树遍历、手写二叉树的遍历、二叉树高度计算，以及二叉树删除，其后给出四个关于树的综合编程实例。图内容主要包括图基本操作、图表示方法，最后给出三个关于图的综合编程实例。

第 5 章递归与分治。主要介绍七个典型的递归与分治问题：汉诺塔、子串组合、数组组合、格子排列、八皇后、循环赛日程安排与棋盘覆盖。

第 6 章集合与映射。主要介绍 STL 集合容器 std::set、映射容器 std::map、多键映射容器 std::multimap，以及哈希映射的用法。同时给出集合相似度与哈希冲突解决的综合实例，

第 7 章 Win32 GUI 编程基础。主要介绍使用 Win32 GUI 开发图形界面程序的基础。在实际工程中，算法的过程或者结果通常都需要可视化，一方面可以用来调试程序，另一方面可以用来展示算法效果。因此本章使用 CodeBlocks 或者 Visual Studio 2015 开发图形界面程序。内容涉及 Win32 消息基础和图形设备接口 GDI。同时给出综合编程实例：简单多边形的创建、绘制、平移与旋转。

第 8 章介绍四个综合编程课程设计实例。分别是扑克洗牌、二叉树重建可视化、L-System 分形树建模，以及迷宫问题。这些实例要求读者能将算法实现和图形界面编程进行结合，从而把最终的算法和交互过程展现出来。

本书由丁展、梁颖红和李广水主编。在编写过程中，我们力求精益求精，但难免存在一些不足之处，如果读者使用本书时遇到问题，可以发送 E-mail 到 dingzh@jit.edu.cn，我们会及时给您回复。

编　者

2018 年 5 月

目　录

第 1 章　字符串处理

字符串几乎在所有程序设计语言中都占据了重要地位。在 C 语言中，C 语言库函数提供了一系列 API 处理各种字符串问题。而在 C++中，可以使用类 std::string、std::istringstream、std::ostringstream，以及若干 STL 算法解决各种字符串问题。

1.1　字符串基本操作

1.1.1　字符串复制

C 语言中可以使用 strcpy 函数实现字符串复制，strcpy 函数声明如下：

```
char * strcpy ( char * destination, const char * source );
```

strcpy 会将 source 指向的字符串内容连同末尾的 null 字符一起复制到 destination 指向的内存空间。为了避免溢出，destination 指向的字符串空间必须能容纳 source 中所有字符(包括 null 字符)，并且 source 和 destination 在内存上不能存在重叠现象。函数将返回 destination 指针。

例如，代码如下：

```
#include <stdio.h>
#include <string.h>

int main ()
{
   char str1[]="Test string";
   char str2[40];
   char str3[40];
   strcpy (str2,str1);
   strcpy (str3,"copy successful");
   printf ("str1: %s\nstr2: %s\nstr3: %s\n",str1,str2,str3);
   return 0;
}
```

输出：

```
str1: Test string
str2: Test string
str3: copy successful
```

一些公司在面试或笔试时喜欢直接使用 strcpy 的内部实现作为题目，以下是一个较为高效的实现过程，代码如下：

```
char *strcpy(char *dest, const char *src)
{
    char *save = dest;
    while(*dest++ = *src++);
    return save;
}
```

除了 strcpy 外，C 语言库函数中还提供了 strncpy 函数用于复制指定数目的字符到目标区域。strncpy 函数声明如下：

```
char * strncpy ( char * destination, const char * source, size_t num );
```

strncpy 从 source 指向的空间中复制前 num 个字符到 destination 指向的空间中。如果还没有复制完 num 个字符时，source 已经达到了字符串尾部，那么函数将补充复制若干个 null 字符到 destination 指向的空间中，直到复制 num 个字符为止。

如果这是第一次碰到 size_t 类型，请别慌张，size_t 是无符号整型，等价于 unsigned int。后续很多函数都会涉及此类型。

strncpy 例子，代码如下：

```
#include <stdio.h>
#include <string.h>

int main ()
{
   char str1[]= "To be or not to be";
   char str2[40];
   char str3[40];

   /* copy to sized buffer (overflow safe): */
   strncpy ( str2, str1, sizeof(str2) );

   /* partial copy (only 5 chars): */
   strncpy ( str3, str2, 5 );
   str3[5] = '\0';    /* null character manually added */

   puts (str1);
   puts (str2);
   puts (str3);

   return 0;
}
```

输出：

```
To be or not to be
To be or not to be
To be
```

需要指出的是，在使用 strncpy 函数时，通常需要自行添加 null 终结符到 destination 中。

在 C++语言中可以直接使用 std::string 类处理各种字符串问题。有三种方法可以对字符串复制：std::string 的构造函数，操作符重载函数 operator=，以及成员函数 assign。

第一种方法：

```
char str1[]= "To be or not to be";
std::string str2 = str1;
std::string str3;
str3 = str1;
```

第二种方法：

```
char str1[]= "To be or not to be";
std::string str2;
str2.assign(str1, str1+strlen(str1));
```

如果只想复制若干个字符，还可以有以下第三种方法：

```
char str1[]= "To be or not to be";
std::string str2;
str2.assign(str1, str1+2);
```

assign 是 std::string 的类成员函数，三个重载函数声明如下：

```
string& assign (const char* s);
string& assign (const char* s, size_t n);
template <class InputIterator>
string& assign (InputIterator first, InputIterator last);
```

1.1.2 字符串连接

C 语言中可以使用 strcat 函数实现字符串的连接，也就是将一个字符串复制到另一个字符串的尾部，strcat 函数声明如下：

```
char * strcat ( char * destination, const char * source );
```

strcpy 会将 source 指向的字符串内容连同末尾的 null 字符一起复制到 destination 末端的内存空间中。在连接开始时，source 的第一个字符将会覆盖 destination 末端的 null 字符，在连接结束时，source 最后的 null 字符将会复制到 destination 的最后。

destination 指向的字符串空间必须能容纳 source，以及 destination 中所有字符数之和（包括 null 字符），并且 source 和 destination 在内存上不能存在重叠现象。函数将返回 destination 指针。

例如，代码如下：

```
#include <stdio.h>
#include <string.h>

int main ()
{
  char str[80];
  strcpy (str,"these ");
  strcat (str,"strings ");
  strcat (str,"are ");
  strcat (str,"concatenated.");
  puts (str);
  return 0;
}
```

输出：

```
these strings are concatenated.
```

一些公司在面试或笔试时也喜欢直接使用 strcat 的内部实现作为题目，以下是一个较为高效的实现过程，代码如下：

```
char *strcat(char *dest, const char *src)
{
    char *save = dest;
    while (*dest)
      dest++;
    while (*dest++ = *src++);
    return save;
}
```

在 C++中，可以直接使用 std::string 类成员函数 append 或者直接调用操作符重载的“+=”符号函数用于字符串的连接。

例如，使用 append 成员函数，代码如下：

```
#include <iostream>
#include <string>

int main ()
{
  std::string str;
  std::string str2="Writing ";
  std::string str3="print 10 and then 5 more";

  // used in the same order as described above:
  str.append(str2);                       // "Writing "
  str.append(str3,6,3);                   // "10 "
  str.append("dots are cool",5);          // "dots "
  str.append("here: ");                   // "here: "
```

```
    str.append(10u,'.');                                   // ".........."
    str.append(str3.begin()+8,str3.end());                 // " and then 5 more"
    str.append<int>(5,0x2E);                               // "....."

    std::cout << str << '\n';
    return 0;
}
```

输出：

```
Writing 10 dots here: ... and then 5 more...
```

或者使用 operator +=成员函数，代码如下：

```
#include <iostream>
#include <string>

int main ()
{
    std::string name ("John");
    std::string family ("Smith");
    name += " K. ";              // c-string
    name += family;              // string
    name += '\n';                // character

    std::cout << name;
    return 0;
}
```

输出：

```
John K. Smith
```

当然，还可以使用全局的 operator +函数来处理 std::string 的连接，代码如下：

```
#include <iostream>
#include <string>

int main ()
{
    std::string firstlevel ("com");
    std::string secondlevel ("google");
    std::string scheme ("http://");
    std::string hostname;
    std::string url;

    hostname = "www." + secondlevel + '.' + firstlevel;
    url = scheme + hostname;
    std::cout << url << '\n';
```

```
    return 0;
}
```

输出：

```
http://www.google.com
```

1.1.3 反转字符串

反转字符串也是一些程序和算法需要的功能。在 Windows 下使用 C 时，可以使用 strrev 反转字符串，strrev 函数声明如下：

```
char * strrev( char *str );
```

例如，代码如下：

```
#include<stdio.h>
#include<string.h>
int main()
{
    char a[] = "my name";
    strrev(a);
    puts(a);
    return   0;
}
```


输出：

```
eman ym
```

需要说明的是，Linux 系统库中并没有包含这个函数，所以必须自己定义反转字符串函数。strrev 的一个简单的实现过程如下，代码如下：

```
#include <string.h>
char *strrev(char *str)
{
    int i = strlen(str) - 1, j = 0;
    char ch;
    while (i > j)
    {
        ch = str[i];
        str[i] = str[j];
        str[j] = ch;
        i--;
        j++;
    }
    return str;
}
```

利用指针和异或运算，可以高效实现字符串的反转，代码如下：

```
void strrev(char *p)
{
    char *q = p;
    while(q && *q) ++q;
    for(--q; p < q; ++p, --q)
        *p = *p ^ *q,
        *q = *p ^ *q,
        *p = *p ^ *q;
}
```

在 C++中，则可以直接使用 STL 算法库中的 std::reverse 函数对字符串或者 std::string 对象进行反转。例如，代码如下：

```
#include <iostream>
#include <string> //for std::string
#include <algorithm> //for std::reverse
#include <cstring> //for strlen
using namespace std;

int main()
{
    std::string s = "Hello world!";
    std::reverse(s.begin(), s.end());
    std::cout<<s<<endl;

    char str[] = "get ready!";
    std::reverse(str, str+strlen(str));
    cout << str << endl;
    return 0;
}
```

输出：

```
!dlrow olleH
!ydaer teg
```

从上例中可以看出，std::reverse 不仅可以对 std::string 对象进行反转，也可以对普通字符串进行反转。std::reverse 是 STL 算法函数之一，所以必须引入头文件<algorithm>。

请大家检查如下字符串反转代码，程序运行时会崩溃，代码如下：

```
#include <iostream>
#include <algorithm> //for std::reverse
#include <cstring> //for strlen
using namespace std;

int main()
```

```
{
    char *str = "get ready!";
    std::reverse(str, str+strlen(str));
    cout << str << endl;
    return 0;
}
```

原因是 str 是一个字符指针，它指向的是一个字符串常量，而常量是无法修改的。

1.1.4 大小写转换

在 C 语言的库函数中，并没有提供对字符串大小写转换的函数。但库函数提供了 toupper 和 tolower 函数用于将单个字符的大小写转换。toupper 将字符转换成大写，tolower 将字符转换成小写。函数声明如下：

```
int toupper ( int c );
int tolower ( int c );
```

如果传入的参数 c 没有对应的大小写字符，那么返回值就是传入的 c 值。另外，这两个库函数声明在<ctype.h>中而不是在<string.h>中。

例如，代码如下：


```
#include <stdio.h>
#include <ctype.h>
int main ()
{
  int i=0;
  char str[]="Test String.\n";
  char c;
  while (str[i])
  {
    c=str[i];
    putchar (tolower(c));
    i++;
  }
  return 0;
}
```

输出：

```
test string.
```

可以利用字符大小写转换的函数，实现字符串的大小写转换，一个基本的小写转换方法如下，代码如下：

```
char* ToLower(char* str)
{
    int i;
```

```
    if(!str) return NULL;
    for(i = 0; str[i]; ++i){
        str[i] = tolower(str[i]);
    }
    return str;
}
```

利用指针进行大写转换的方法如下，代码如下：

```
char* ToUpper(char *str)
{
    char* ostr = str;
    if(!str) return NULL;
    for ( ; *str; ++str) *str = toupper(*str);
    return ostr;
}
```

在 C++中，可以直接使用 STL 算法库中的 std::transform 函数对字符串或者 std::string 对象进行大小写转换。代码如下：

```
#include <iostream>
#include <algorithm> //for std::transform
#include <cstring> //for strlen
#include <string> //for std::string

using namespace std;
int main()
{
    std::string str = "Hello World";
    std::transform(str.begin(), str.end(), str.begin(), ::toupper);
    cout<<str<<endl;

    char str2[] = "My Name";
    std::transform(str2, str2+strlen(str2), str2, ::tolower);
    cout<<str2<<endl;
}
```

输出：

```
HELLO WORLD
my name
```

这里 std::transform 函数的声明如下：

```
template <class InputIterator, class OutputIterator, class UnaryOperation>
  OutputIterator transform (InputIterator first1, InputIterator last1,
                            OutputIterator result, UnaryOperation op);
```

其中，第一参数 first1 和第二参数 last1 比较容易理解，它们代表输入的起始和终止迭代器位置。而第三个参数是进行变换后数据要输出的迭代器位置。如果想把这个变换函数应用到容器自身，那么只要将 result 的值设置成和 first1 一样就可以。最后一个参数是一元变换函数。

如果这是你第一次接触迭代器，那么不用慌张，可以把它想象成指针。后面章节中还会出现各种迭代器的用法。

1.1.5 字符串与数的转换

首先，学习 C 语言中字符串与数字的互换。在 C 语言中，sprintf 函数可以将任何一个或多个数字格式化成指定格式的字符串。sprintf 函数声明如下：

```
int sprintf ( char * str, const char * format, ... );
```

和 printf 函数相比，它多了第一个参数 str。这是因为 printf 函数只需要将格式化的内容输出到 console 窗口，也就是屏幕。而 sprintf 需要将格式化的内容输出到缓冲区，第一个参数 str 就是这个缓冲区内存。因为是输出到缓冲区，所以需要自行确保 str 的内存空间足够容纳格式化的内容，否则存在内存溢出风险。最后需要指出的是，一个 null 字符会被自动添加到格式化内容的最后位置。

例如，代码如下：

```
#include <stdio.h>

int main ()
{
  char buffer [64];
  int n, a=6, b=8;
  n=sprintf (buffer, "%d plus %d is %d", a, b, a+b);
  printf ("[%s] is a string %d chars long\n",buffer,n);
  return 0;
}
```

输出：

```
[6 plus 8 is 14] is a string 14 chars long
```

和 printf 一样，如果想以指定精度输出浮点数，参照如下代码：

```
double a = 3.1415926;
sprintf(buffer, “%.2f”, a);
```

输出：

```
3.14
```

在 C 语言中，将字符串变成一个整数，可以使用 atoi 函数，需要引入头文件<stdlib.h>。atoi 函数声明如下：

```
int atoi (const char * str);
```

例如，代码如下：

```
#include <stdio.h>          /* printf, fgets */
#include <stdlib.h>         /* atoi */

int main ()
{
  int i;
  char buffer[256];
  printf ("Enter a number: ");
  fgets (buffer, 256, stdin);
  i = atoi (buffer);
  printf ("The value entered is %d.",i);
  return 0;
}
```

输出：

```
Enter a number: 64
The value entered is 64.
```

将字符串变成一个浮点数，可以使用 atof 函数。atof 函数声明如下：

```
double atof (const char* str);
```

例如，代码如下：

```
#include <stdio.h>          /* printf, fgets */
#include <stdlib.h>         /* atof */
#include <math.h>           /* sin */

int main ()
{
  double n,m;
  double pi=3.1415926535;
  char buffer[256];
  printf ("Enter degrees: ");
  fgets (buffer,256,stdin);
  n = atof (buffer);
  m = sin (n*pi/180);
  printf ("The sine of %f degrees is %f\n" , n, m);
  return 0;
}
```

输出：

```
Enter degrees: 45
The sine of 45.000000 degrees is 0.707101
```

相对于 atoi 和 atof，C 语言还提供了 sscanf 函数用于从格式化的字符串中提取多个类型的数据。从名字上可以看出 sscanf 与 scanf 的作用类似，区别在于 scanf 是从用户键盘输入的字符中提取数据，而 sscanf 是从一个字符串缓冲中提取数据。sscanf 函数声明如下：

```
int sscanf ( const char * s, const char * format, ...);
```

例如，代码如下：

```
#include <stdio.h>

int main ()
{
   char str[]="I am 18 years old";
   char subStr[20];
   int n;

   sscanf (str,"%s %*s %d",subStr,&n);
   printf ("%s -> %d\n",subStr, n);

   return 0;
}
```

输出：

```
I -> 18
```


在 C++中，新手通常认为 std::string 类应该提供字符串与数字转换的成员函数，但实际上 std::string 类并没有提供任何字符串与数字转换的方法，这个任务交给了 stringstream 流。

想把字符串转换成数字，使用 istringstream 类，需要引入头文件<sstream>。 例如，代码如下：

```
#include <iostream>
#include <sstream> //std::istringstream
#include <string>   //std::string

using namespace std;
int main ()
{
   char str[]="3.5 + 2.5 = 6";
   std::istringstream iss(str);
   double a, b, c;
   std::string s;

   iss>>a>>s>>b>>s>>c;
   cout<<"a = "<<a<<", b = "<<b<<", c = "<<c<<endl;
   return 0;
}
```

输出：

```
a = 3.5, b = 2.5, c = 6
```

想把数字转换成字符串，使用 ostringstream 类，同样需要引入头文件<sstream>。 例如，代码如下：

```
#include <iostream>
#include <sstream> //std::ostringstream
#include <string>    //std::string

using namespace std;
int main ()
{
   std::ostringstream oss;
   double a=3.5, b=2.5, c;
   c = a+b;
   std::string s;

   oss<<a<<" + "<<b<<" = "<<c;
   s = oss.str();
   cout<<s<<endl;
   return 0;
}
```

输出：

```
3.5 + 2.5 = 6
```

如果编译器完全支持 C++11 标准，那么还可以使用新 STL 标准的库函数 std::stoi, std::stof 和 std::stod 从 std::string 对象获得 int, float 和 double 值。三个函数声明如下：

```
int stoi (const string&    str, size_t* idx = 0, int base = 10);
float stof (const string&    str, size_t* idx = 0);
double stod (const string&    str, size_t* idx = 0);
```

例如，代码如下：

```
#include <iostream>       // std::cout
#include <string>         // std::string, std::stoi, std::stod

int main()
{
      std::string str = "1997, my";
      std::string dataStr("3.1415 6.2830");
      size_t sz;
      int year = std::stoi(str);
      double pi = std::stod(dataStr, &sz);
      double pi2 = std::stod(dataStr.substr(sz));
      std::cout << year << " " << pi << " " << pi2<<std::endl;
      return 0;
}
```

输出：

```
1997 3.1415 6.283
```

1.1.6 字符串查找

在 C 语言的库函数中，strstr 函数用于从一个字符串中查找一个子串，函数声明如下：

```
char * strstr ( const char * str1, const char * str2 );
```

函数会返回 str2 在 str1 中第一次出现的位置。如果 str1 中没有 str2 出现，则返回 null 指针。例如，代码如下：

```
#include <stdio.h>
#include <string.h>

int main ()
{
   char str[] ="It is apple tree.";
   char * pch;
   pch = strstr (str,"apple");
   strncpy (pch,"lemon",5);
   puts (str);
   return 0;
}
```

输出：

```
It is lemon tree.
```

如果只想从字符串中查找某个字符出现的位置，则可以使用 strchr 函数，声明如下：

```
char * strchr ( const char * str, int c);
```

例如，代码如下：

```
#include <stdio.h>
#include <string.h>

int main ()
{
   char str[] = "It's a long long story.";
   char * pch;
   printf ("Looking for the 'l' character in \"%s\"...\n",str);
   pch=strchr(str,'l');
   while (pch!=NULL)
   {
      printf ("found at %d\n",pch-str+1);
      pch=strchr(pch+1,'l');
```

```
    }
    return 0;
}
```

输出：

```
Looking for the 'l' character in "It's a long long story."...
found at 8
found at 13
```

如果想从字符串中查找任意一个指定的字符出现的位置，则可以使用 strpbrk 函数。读者可以自行搜索该函数的用法，这里不再赘述。

在 C++中，std::string 类成员函数 find 可以用来查找指定的字符串或字符，四个 find 重载函数声明如下：

```
size_t find (const string& str, size_t pos = 0) const;
size_t find (const char* s, size_t pos = 0) const;
size_t find (const char* s, size_t pos, size_t n) const;
size_t find (char c, size_t pos = 0) const;
```

例如，代码如下：

```
#include <iostream>              // std::cout
#include <string>                // std::string

int main ()
{
  std::string str ("Gather as many roses as you possibly can.");
  std::size_t found = str.find("as");
  while (found!=std::string::npos)
  {
    std::cout << "found at: " << found << '\n';
    found = str.find("as", found+1);
  }
  return 0;
}
```

输出：

```
found at: 7
found at: 21
```

需要指出的是，在 std::string 对象中如果没有找到指定的字符或字符串，其返回的 size_t 的值是 std::string::npos。使用如下的代码，就能明白 std::string::npos 的具体值。

```
std::cout<<std::hex<<std::string::npos<<std::endl;
```

答案就是：

```
ffffffff
```

也就是(unsigned int)-1。

如果想从 std::string 中查找指定的字符串中任意一个字符出现的位置，则可以使用成员函数 find_first_of 或者 find_last_of 函数。读者可以自行搜索该函数的用法。

1.1.7 删除字符

C 语言的库函数中并没有提供删除字符串中指定字符的函数。以下是一个较为高效的字符删除算法，代码如下：

```
char* remchr(char *str, char c)
{
    char *src, *dst;
    for (src = dst = str; *src != '\0'; src++) {
        *dst = *src;
        if (*dst != c) dst++;
    }
    *dst = '\0';
    return str;
}
```

算法的巧妙之处在于 dst 偏移量始终不会超过 src 指针的偏移量，所以避免了额外开辟内存空间用于存储新的字符串。利用该方法举例，代码如下：


```
#include <stdio.h>
char* remchr(char *str, char c)
{
    char *src, *dst;
    for (src = dst = str; *src != '\0'; src++) {
        *dst = *src;
        if (*dst != c) dst++;
    }
    *dst = '\0';
    return str;
}

int main ()
{
  char str[]="test string.\n";
  puts(remchr(remchr(str, 't'), 's'));
  return 0;
}
```

输出：

```
e ring.
```

在 C++中，std::string 类没有提供去除指定字符的成员函数，这的确令人沮丧。但是，STL

的算法库提供了一个 std::remove 函数，它可以删除两个迭代器区间内的指定值。例如，代码如下：

```
#include <iostream>          // std::cout
#include <string>            // std::string
#include <algorithm>         // std::remove

int main ()
{
    std::string str ("Gather as many roses as you can.");
    std::remove(str.begin(), str.end(), 'a');
    std::cout<<str;
    return 0;
}
```

输出：

```
Gther s mny roses s you cn. can.
```

以上输出结果中，最后五个字符(含一个空格) can.是怎么回事？事实上，std::remove 只负责从区间内删除指定的值，但它不负责调整（缩小）容器的容量。此时，std::remove 的返回值可以提供一些帮助。它的返回值是一个迭代器，指向新区间的末端位置。因此最终的方法就是结合 std::remove 和 std::string 类成员函数 erase，实现指定字符的删除功能。标准写法如下：

```
str.erase(std::remove(str.begin(), str.end(), 'a'), str.end());
```

例如，代码如下：

```
#include <iostream>          // std::cout
#include <string>            // std::string
#include <algorithm>

int main ()
{
    std::string str ("Gather as many roses as you can.");
    str.erase(std::remove(str.begin(), str.end(), 'a'), str.end());
    std::cout<<str;
    return 0;
}
```

输出：

```
Gther s mny roses s you cn.
```

1.1.8 字符串字典比较

在 C 语言的库函数中，提供了 strcmp 函数用于比较两个字符串的字典顺序。函数声明如下：

```
int strcmp ( const char * str1, const char * str2 );
```

strcmp 返回值见表 1-1。

表 1-1　strcmp 返回值

返回值	含义
<0	str1 在字典中排在 str2 的前面
=0	str1 和 str2 完全相同
>0	str1 在字典中排在 str2 的后面

例如，代码如下：

```
#include <stdio.h>
#include <string.h>

int main ()
{
  char key[] = "volleyball";
  char buffer[80];
  do {
    printf ("Guess my favorite sport? ");
    fflush (stdout);
    scanf ("%79s",buffer);
  } while (strcmp (key,buffer) != 0);
  puts ("Correct answer!");
  return 0;
}
```

输出：

```
Guess my favorite sport? football
Guess my favorite sport? basketball
Guess my favorite sport? volleyball
Correct answer!
```

C 语言库函数中还提供了 strncmp 函数用于比较两个字符串前面若干字节的字典顺序。函数声明如下：

```
int strncmp ( const char * str1, const char * str2, size_t num );
```

可以看到，strncmp 比 strcmp 多了最后一个参数，它表示要比较的字节长度。strncmp 用法如下，代码如下：

```
#include <stdio.h>
#include <string.h>

int main ()
{
  char str[][10] = { "Green" , "Apple" , "Great", "Ten" };
  int n;
  puts ("Looking for string starting with \"Gr\"...");
```

```
    for (n=0 ; n<4 ; n++)
    {
      if (strncmp (str[n],"Gr",2) == 0)
      {
        printf ("found %s\n",str[n]);
      }
    }
    return 0;
}
```

输出：

```
Looking for string starting with "Gr"...
found Green
found Great
```

需要指出的是，strcmp 和 strncmp 都是对大小写敏感的字符串比较函数，这也意味"apple"和"Apple"在字典中不是相同的字符串。如果调用 strcmp("apple", "Apple")，它的返回值将是一个正值，这时因为在 ASCII 码表中，字符'a'大于'A'。

那么如何不区分大小写，进行实际意义的字典序比较呢？C 语言库函数中，提供了库函数解决这个问题。需要指出的是，在 Windows、Linux 和 Mac 下，他们使用了不同的库函数名。在 Windows 下，这个函数叫作 stricmp，但在 Linux 或者 Mac 下，它叫作 strcasecmp。所以，如果你是一个跨平台的 C/C++程序员，通常需要预先通过宏定义的方式统一这个函数的调用，方法如下：

```
#if  defined (LINUX) || defined (_MACX)
   #define _stricmp strcasecmp
   #define stricmp strcasecmp
#endif
```

关于 strcmp 函数，最后还要强调两点：

（1）strcmp 的返回值为 0 才表示两个字符串是完全一样的；

（2）为了检测 str1 在字典中排在 str2 的前面，不要写成 strcmp(str1, str2)==-1，而应该更稳妥地写成 strcmp(str1, str2)<0。

在 C++中，std::string 类的字典序比较功能非常强大，它提供了操作符重载的成员函数 <, <=, ==，!=, >, >=进行字典序比较。举例如下，代码如下：

```
#include <iostream>
#include <string>

int main ()
{
   std::string a = "alpha";
   std::string b = "beta";

   if (a==b) std::cout << "a and b are equal\n";
   if (a!=b) std::cout << "a and b are not equal\n";
```

```
    if (a< b) std::cout << "a is less than b\n";
    if (a> b) std::cout << "a is greater than b\n";
    if (a<=b) std::cout << "a is less than or equal to b\n";
    if (a>=b) std::cout << "a is greater than or equal to b\n";
    return 0;
}
```

输出：

```
a and b are not equal
a is less than b
a is less than or equal to b
```

不过，std::string 类中并没有提供不区分大小写的字典序比较函数。所以，在调用操作符重载的比较函数之前，可以先调用 std::transform 进行小写转换：

```
#include <algorithm>
...
std::transform(a.begin(), a.end(), a.begin(), tolower);
std::transform(b.begin(), b.end(), b.begin(), tolower);
```

1.2 字符串处理常见问题

1.2.1 居民身份证号的表示

中国居民身份证号的存储是程序设计中经常碰到的问题。当今，中国居民的身份证号码有 15 位和 18 位两种。1985 年我国实行居民身份证制度，当时签发的身份证号码是 15 位，而 1999 年签发的身份证由于年份的扩展和末尾加了校验码，就成了 18 位。我们知道一个 int 或者 long 类型最多只能表达 20 多亿的数字，也就是 10 位整数，这远远不够表达一个身份证号。幸好还有一个 long long 类型，可以通过如下代码知道它的最大值，代码如下：

```
#include <iostream>
#include <limits>
int main ()
{
    std::cout<<std::numeric_limits<long long>::max();
    return 0;
}
```

输出：

```
9223372036854775807
```

这个最大值一共是 19 位，已经足够容纳中国居民身份证号的 18 位。似乎用 long long 问题就能解决。不过可惜的是，有些人的身份证号的最后一位是字母 X 而不是数字。这是因为，18 位身份证号的最后一位作为尾号的校验码。它是由号码编制单位按统一的公式计算的，如果某人的尾号是0～9，就不会出现X；但如果尾号恰好是10，那么就得用X来代替，因为如果用10作为尾号，那么此人的身份证就变成了19位，这不符合国家标准。由于X符号是罗马数字的10，所以干脆用字母X来代替X，即表示10。基于上述事实，这里只能用字符数组表示身份证号码：

```
char ID[19];
```

这里多设置一位，是为结尾的null字符准备的。

1.2.2 std::string与const char*转换

程序有时需要得到一个std::string对象的字符串表示形式。std::string类提供了c_str函数解决这个问题，代码如下：

```
#include <iostream>     // std::cout
#include <string>       // std::string
#include <cstdlib>      //atoi
int main ()
{
   std::string str("365");
   int year = atoi(str.c_str());
   std::cout<<year<<std::endl;
   return 0;
}
```

调用 atoi 函数时需要传入一个常量字符串，所以使用成员函数 str.c_str()就能获得std::string对象的const char*表示形式。

将const char*转换成std::string，那么这更简单，因为std::string其中一个构造函数就是以const char*作为参数。

1.2.3 字符串与buffer缓冲

作为新手程序员来说，往往认为字符串和buffer缓冲是同一个含义。但实际上，字符串就是一串以null作为终结符的字符，而buffer缓冲没有null终结符的概念，它的尾部不需要也不在乎是否有null终结符。新手之所以会混淆字符串和buffer缓冲，是因为定义buffer时候程序通常将其定义为char[]或者char*类型。所以这里可以通过定义一个byte类型变量，区分char和byte。在win32系统API中，这是区分字符串和butter的常用做法。示例代码如下：

```
#include <iostream>     // std::cout
#include <cstring>      // strcmp, memcpy

typedef char BYTE;

int main ()
{
      char * ID   = new char[19];
```

```
    strcpy(ID, "320118196706210599");

    BYTE * buffer   = new BYTE[18];
    memcpy(buffer, ID, sizeof(BYTE)*18);

    std::cout<<ID<<std::endl;
    std::cout.write(buffer, 18);
    std::cout<<std::endl;
    //don't try to cout a buffer using stream method.
    //std::cout<<buffer<<std::endl;

    delete[] ID;
    delete[] buffer;
    return 0;
}
```

上述代码中，ID 是一个存储身份证号码的字符串，所以需要 19 个字节空间，其中最后一个字节用于存储 null 终结符。而 buffer 只是一块缓冲区，它只要存储身份证号的所有内容就可以，所以它不需要额外的 null 终结符。

1.2.4 设置浮点数精度

在各种在线程序评测系统中（如 PAT、PTA、ZOJ 等），按指定浮点精度输出浮点数是常见的要求，所以该项技能必须掌握。

对于 C 语言，方法很简单，写成形如"%.2f"的形式（2 位浮点精度）。

```
printf("%.2f", 3.1415926);
```

再加上指定宽度限制，如 6 字符宽度，2 位浮点精度。

```
printf("%6.2f", 3.1415926);
```

如果希望左对齐。

```
printf("%-6.2f\n", 3.1415926);
```

对于 C++语言，2 位浮点精度输出方法如下。

```
std::cout<<std::fixed<<std::setprecision(2)<<3.1415926<<std::endl;
```

注意加上头文件<iomanip>，其中 manip 是 manipulator 的简写，即操作符。

如果加上指定宽度限制，如 6 字符宽度，2 位浮点精度。

```
std::cout << std::fixed << std::setw(6) << std::setprecision(2) << 3.1415926 << std::endl;
```

如果希望左对齐。

```
std::cout << std::fixed <<std::left<< std::setw(6) << std::setprecision(2) << 3.1415926 << std::endl;
```

1.2.5 得到一行输入的字符串

在 C 语言中，要得到一行用户输入的字符串，首先想到的可能是 gets 函数。没错，gets 的确可以满足各种竞赛程序的要求。但是，gets 函数的最大问题就是容易出现字符串溢出问题，它的函数参数中并没有字符串长度的限制。所以，从 C11 及 C++11 标准后，gets 已经从标准中删除。正确做法就是用 fgets 替代 gets。fgets 函数声明如下：

```
char * fgets ( char * str, int num, FILE * stream );
```

因为标准输入等价于一种文件流，所以最后一个参数用 stdin 就可以从键盘中得到一行字符串，示例代码如下：

```
#include <stdio.h>

int main ()
{
    char str[64];
    fgets(str, 64, stdin);
    puts(str);
    return 0;
}
```

尤其需要注意的是，fgets 会将回车符也一并发送到字符串末尾，之后再添加一个 null 终结符。

在 C++中，有两种方法可以得到一行输入的字符串。

（1）使用 cin 的成员函数 getline，示例如下：

```
char buffer[64];
std::cin.getline(buffer, 64);
```

（2）使用 STL 库中的全局函数 std::getline，示例如下：

```
std::string str;
std::getline(std::cin, str);
```

方法（1）需要指定输入字符串的最大长度，而方法（2）没有这个限制。所以在实际开发中，作者更倾向于使用方法（2）。

示例代码如下：

```
#include <iostream>

int main ()
{
    char buffer[64];
    std::cin.getline(buffer, 64);
    std::cout<<buffer<<std::endl;

    std::string str;
```

```
    std::getline(std::cin, str);
    std::cout<<str<<std::endl;
    return 0;
}
```

1.2.6 统计一行文本中单词个数

统计一行文本中单词的个数是一个常见的字符串处理问题。这里“单词”是指连续不含空格的字符串，各单词之间可以用空格、制表符等分隔，空格数或制表符个数可以是连续的多个。

本题最直观的解法就是逐个读取字符，然后分析空格、制表符，以及最后的回车键。处理空格和制表符时要注意连续出现的情形。这里使用 C 和 C++提供两种较为通用的解法。

C 语言方法的代码如下：

```
#include <stdio.h>

int main()
{
    char str[1024], word[1024];
    int count = 0, pos=0, curPos=0;
    fgets(str, 1024, stdin);

    while(sscanf(str+curPos, "%s%n", word, &pos)!=EOF)
    {
        ++count;
        curPos+=pos;
    }

    printf("%d\n", count);
    return 0;
}
```

该方法首先用 fgets 获取一行文本，然后在 sscanf 调用时使用了%n 格式化符号。%n 并不是表示和回车相关的信息，它表示 sscanf 函数从字符串 str 中已经读取的字节数，也就是当前读到哪里。要读取下一个单词，只需要从 str+curPos 开始下一次的读取。

C++语言方法的代码如下：

```
#include <iostream>
#include <string>
#include <sstream>

using namespace std;
int main()
{
```

```
        std::string str;
        std::istringstream iss;
        int num=0;

        std::getline(cin, str);
        iss.str(str);
        while(iss>>str)
        {
            ++num;
        }
        cout<<num<<endl;
        return 0;
}
```

首先用 std::getline 获取一行文本，接着用 std::istringstream 获得每个单词。

1.2.7 std::stream 的高速缓冲方法

以下是一段很简单的 C 语言代码，计算机使用 Release 版本，它执行了 6 秒钟，代码如下：

```
#include <stdio.h>
int main ()
{
  int i;
  for(i=0; i<100000; ++i)
    printf("hello, world!\n");
  return 0;
}
```

将上述代码换成 C++的常用写法，代码如下：

```
#include <iostream>

int main ()
{
    int i;
    for(i=0; i<100000; ++i)
        std::cout<<"hello, world!\n";
    return 0;
}
```

同样使用 Release 版本程序，执行了大约 9 秒钟，即比 C 语言程序慢了约 50%。这并不是偶然现象，但并不是所有 C++程序都比同样功能的 C 语言程序要慢。那问题在哪里？答案是 cout 默认采用的同步方式。默认情况下，iostream 对象要和 cstdio 流同步。“同步”这个词在操作系统并发、多线程程序设计等处频繁出现。在这里，同步的意思是 C++程序中可能在某些地方使用了 std::cin 或 std::cout，而在别的地方又使用了 scanf 或者 printf。iosteram 对象

（std::cin 和 std::cout）在默认情况下和 cstdio 流进行同步，也就是确保混合使用 std::cin/std::cout 和 scanf/printf 时程序能正确运行。这种同步以牺牲 std::cin/std::cout 性能为代价。解决方法就是禁止这种同步现象。利用 std::ios 的静态函数 sync_with_stdio(false)，代码如下：

```
#include <iostream>

int main ()
{
    int i;
    std::ios::sync_with_stdio(false); //stop synchronize
    for(i=0; i<100000; ++i)
        std::cout<<"hello, world!\n";
    return 0;
}
```

同样使用 Release 版本程序，以上程序执行了 6 秒钟，和 C 语言代码一样。添加一句提示代码，修改代码如下：

```
#include <iostream>

int main ()
{
    int i;
    std::cout<<"Please try my new method!\n";
    std::ios::sync_with_stdio(false);
    for(i=0; i<100000; ++i)
        std::cout<<"hello, world!\n";
    return 0;
}
```

同样使用 Release 版本程序，以上程序也执行了大约 6 秒钟，符合预期。但是，语句 std::cout<<"Please try my new method!\n";实际上是一颗埋藏的地雷。再读一下上面的代码，它的意思是"Please try my new method!\n"和 cstdio 同步， 而 100 000 遍的"hello, world!\n"和 cstdio 异步。实际上，iostream 和 cstdio 的同步或者异步是在第一次调用 cin 或者 cout 时就决定的，也就是 100000 遍的"hello, world!\n"在理论上采用的还是和 cstdio 同步，std::ios::sync_with_stdio(false);是没法取消已有的同步的。你也许会反驳“两次时间不都是 6 秒吗？不都比 9 秒快吗？”。没错，但不是每次都会有这么好的运气。所以，正确的代码应该如下：

```
#include <iostream>

int main ()
{
    std::ios::sync_with_stdio(false);

    int i;
    std::cout<<"Please try my new method!\n";
```

```
        for(i=0; i<100000; ++i)
            std::cout<<"hello, world!\n";
        return 0;
    }
```

就是把 std::ios::sync_with_stdio(false);语句放在 main 函数内部的第一行。

最后需要强调的是，一旦决定放弃 iostream 和 cstdio 的同步，就千万不要尝试混合使用 printf/scanf 和 std::cin/std::cout。

1.3 综合编程实例：公民数据模拟

公民信息属于个人隐私范围，受我国法律保护。如果软件公司需要大量公民数据进行软件测试，唯一合法方法就是使用模拟的公民数据。因此，本实例要求读者能模拟生成批量的公民数据，模拟数据越真实对软件测试帮助越大。

一个中国公民的信息通常包括：姓名、出生年月、性别、籍贯、身份证号、户口所在地、婚姻状况、宗教信仰、手机号码、工作单位等。为了简化，这里只生成其中部分数据。

1. 居民身份证号编码规则

居民身份证号码是由十七位数字码和一位校验码组成的。排列顺序从左至右依次为：六位数字地址码，八位数字出生日期码，三位数字顺序码和一位校验码，可以用字母表示为 ABCDEFYYYYMMDDXXXR。其含义如下。

（1）地址码（ABCDEF）：表示编码对象常住户口所在县(市、旗、区)的行政区划代码，按 GB/T2260 的规定执行。

（2）出生日期码（YYYYMMDD）：表示编码对象出生的年、月、日。

（3）顺序码（XXX）：表示在同一地址码所标识的区域范围内，对同年、同月、同日出生的人编定的顺序号，顺序码奇数表示男性，偶数为女性。

（4）校验码（R），一位数字，通过前 17 位数字计算得出。

2. 地址码

地址码具体省（直辖市，自治区，特别行政区）代码 AB 如下：

- 11-15 京 津 冀 晋 蒙
- 21-23 辽 吉 黑
- 31-37 沪 苏 浙 皖 闽 赣 鲁
- 41-46 豫 鄂 湘 粤 桂 琼
- 50-54 渝 川 贵 云 藏
- 61-65 陕 甘 青 宁 新
- 81-82 港 澳

CD 是具体省（直辖市，自治区，特别行政区）的城市代码，例如，2101 表示沈阳，2102 表示大连。EF 是市辖区、郊区、郊县、县级市代码。

实际上只有以上这些规则很难生成具体的地址信息，读者可以通过网址（http://www.stats.

gov.cn/tjsj/tjbz/tjyqhdmhcxhfdm/2016/index.html）获得2016年统计用区域代码和城乡划分代码：

例如，江苏省代码为32，江苏省内各城市代码如下：

3201 南京市
3202 无锡市
3203 徐州市
3204 常州市
3205 苏州市
3206 南通市
3207 连云港市
3208 淮安市
3209 盐城市
3210 扬州市
3211 镇江市
3212 泰州市
3213 宿迁市

以南京市为例，区域代码如下：

320100 南京市
320102 玄武区
320104 秦淮区
320105 建邺区
320106 鼓楼区
320111 浦口区
320113 栖霞区
320114 雨花台区
320115 江宁区
320116 六合区
320117 溧水区
320118 高淳区

基于上述信息，如果我们想随机生成某个省、市、区的具体代码信息，首先需要将这些标准信息写入一个数据库或者简单的数组中，以备随机抽取。

3. 顺序码和校验码

顺序码（XXX）（身份证第十五位到十七位）是县、区级政府所辖派出所的分配码，其中单数为男性分配码，双数为女性分配码，例如，003 就是男性。

校验码（R）是通过一系列数学计算得出的，计算方法如下：

首先对前 17 位数字本体码加权求和，其中各位数字对应的权值依次为： 7 9 10 5 8 4 2 1 6 3 7 9 10 5 8 4 2；算出加权求和值后，以 11 对计算结果取模，余数从 0 到 10。各个余数对应的校验码是：1 0 X 9 8 7 6 5 4 3 2。如果校验码不符合这个规则，则肯定是假号码。

关于 18 位身份证号码尾数是“X”的一种解释：因为按照上面的规则，校验码有 11 个，而不是 10 个，所以不能用 0～9 表示。所以如果尾号是 10，那么就得用 X 来代替，X 是罗马数字的 10，用 X 来代替 10，可以保证公民的身份证符合国家标准。

4. 代码实现

先实现姓名的模拟。首先构建一组中国常用姓的数组 familyNames、一组男生常用名字数组 boyNames，以及一组女生常用名字数组 girlNames。然后采用随机函数从 familyNames 中首先抽取一个姓，之后给定一个性别，就可以从男生或者女生常用名字数组中抽取一个名字。NameGenerator 代码如下：

```
void NameGenerator(Gender gender, char name[64])
{
    static const char *familyNames[] = { "赵","钱","孙","李","周","吴","郑","王","冯","陈","褚","卫",
....};
    static const char *boyNames[]={"苑博","伟泽", "熠彤","鸿煊","博涛","烨霖","烨华","煜祺",
... };

    static const char *girlNames[] = {"莹","雪","琳","晗","涵","琴","晴","丽","美","瑶","梦","茜","倩",
...};
    strcpy(name, familyNames[rand()%numFamilyNames]);
    if(gender==MALE)
        strcat(name, boyNames[rand()%numBoyNames]);
    else
        strcat(name, girlNames[rand()%numGirlNames]);
}
```

再实现出生日的模拟。出生年、月、日按顺序单独随机生成。年，需要随机生成在一个指定区间，月则在[1,12]内。而日的取值范围需要根据月生成，尤其要注意闰年情况，代码如下：

```
void BirthdayGenerator(char birthday[9])
{
    int yearStart=1949, yearEnd = 2018;
    int year = rand()%(yearEnd-yearStart)+yearStart;
    int month = rand()%12+1;
    bool bLeapYear = (year%4==0 && year%100!=0)||(year%400==0);
    int days = 0;
    switch(month)
    {
    case 1:
    case 3:
    case 5:
    case 7:
    case 8:
    case 10:
    case 12:
        days = 31;
```

```
            break;
        case 2:
            days = bLeapYear?29:28;
            break;
        default:
            days = 30;
            break;
        }
        days = rand()%days+1;
        sprintf(birthday, "%04d%02d%02d", year, month, days);
    }
```

然后实现身份证中出生地和对应编码号的模拟，这也是整个程序中最重要的部分。首先，我们做了简化，只抽取了几个省中几个市的编码到 cityDatabase.dat 数据库中。这是一种简单的数据表示方式，如果地区编码只有两个数字，这表示省或直辖市；如果地区编码有四个数字，则表示当前省中的某个城市；如果地区编码有六个数字，则表示当前城市中的某个区县。根据这种格式，样例代码定义了一个数据库载入并解析的函数 LoadDatabase，代码如下：

```
bool RegionDatabase::LoadDatabase(const std::string&filename)
{
    std::ifstream file;
    file.open(filename.c_str());
    if(file.fail())
        return false;
    std::string str, code, name;
    CityData currentCity;
    ProvinceData currentProvince;
    bool bCityValid=false;
    std::istringstream iss;
    NameCode nameCode;
    while(true)
    {
        std::getline(file, str);
        if(file.eof())
            break;
        if(file.fail())
            break;
        iss.clear();
        iss.str(str);
        //读入区域码和名称
        iss>>code>>name;
        if(iss.fail())
            continue;
        if(code.length()==2) //读入新的省数据
```

```
            {
                //将上一个省数据打包
                if(currentCity.districtCode.size())
                    currentProvince.m_CityData.push_back(currentCity);
                if(currentProvince.m_CityData.size())
                    m_ProvinceData.push_back(currentProvince);
                currentProvince.m_CityData.clear();
                strcpy(currentProvince.code.code, code.c_str());
                strcpy(currentProvince.code.name, name.c_str());
                bCityValid = false;
            }
            else if(code.length()==4) //读入新的市数据
            {
                //将上一个市数据打包
                if(bCityValid)
                    currentProvince.m_CityData.push_back(currentCity);
                strcpy(currentCity.code.code, code.c_str()+2) ;
                strcpy(currentCity.code.name, name.c_str()) ;
                currentCity.districtCode.clear();
                bCityValid = true;
            }
            else if(code.length()==6) //读入新的区数据
            {
                strcpy(nameCode.name, name.c_str());
                strcpy(nameCode.code, code.c_str()+4);
                currentCity.districtCode.push_back(nameCode);
            }
        }
        //打包最后一个省的数据
        if(currentCity.districtCode.size())
        {
            currentProvince.m_CityData.push_back(currentCity);
            m_ProvinceData.push_back(currentProvince);
        }
        file.close();
        return true;
    }
```

载入数据库后，就可以从中随机抽取地区编码和名称，代码如下：

```
void RegionDatabase::RandomOne(char code[7], char name[128])
{
    code[0] = '\0';
    name[0] = '\0';
    int provinceID = rand()%m_ProvinceData.size();
```

```
    const ProvinceData& provinceData = m_ProvinceData[provinceID];
    strcat(code, provinceData.code.code);
    strcat(name, provinceData.code.name);
    int cityID = rand()%provinceData.m_CityData.size();
    const CityData& cityData = provinceData.m_CityData[cityID];
    strcat(code, cityData.code.code);
    if(strcmp(name, cityData.code.name))
        strcat(name, cityData.code.name);
    int districtID = rand()%cityData.districtCode.size();
    strcat(code, cityData.districtCode[districtID].code);
    strcat(name, cityData.districtCode[districtID].name);
}
```

当名字、出生年月、身份证中出生地及编码都可以随机生成后，下面就要将上述内容组合，代码如下：

```
void CitizenshipSimulator::RandomOne(Citizenship& citizen)
{
    char birthday[9], name[64], regionCode[7], birthOrderStr[4];
    Gender gender;
    int birthOrder = rand()%900;

    citizen.gender = (rand()%2)?MALE:FEMALE;
    NameGenerator(citizen.gender, citizen.name);
    BirthdayGenerator(birthday);
    m_RegionDatabase.RandomOne(regionCode, citizen.region);
    strcpy(citizen.ID, regionCode);
    strcat(citizen.ID, birthday);
    sprintf(birthOrderStr, "%03d", birthOrder);
    strcat(citizen.ID, birthOrderStr);
    VerifyCode(citizen.ID);
}
```

身份证号最后一位校验码代码如下：

```
void CitizenshipSimulator::VerifyCode(char ID[19])
{
    static int weight[17]={7, 9, 10, 5, 8, 4, 2, 1, 6, 3, 7, 9, 10, 5, 8, 4, 2};
    static char crc[12] = "10X98765432";
    int code = 0, i;
    for(i=0; i<17; ++i)
        code+=weight[i]*(ID[i]-'0');
    code %=11;
    ID[17] = crc[code];
}
```

下面是一个模拟结果：

```
顾楷瑞      MALE      110106197510073826      北京丰台区
澹台修杰    MALE      13013119761124652X      河北省石家庄市平山县
时皎月      FEMALE    130107196907041940      河北省石家庄市井陉矿区
毛娇然      FEMALE    120116199402244572      天津滨海新区
花立轩      MALE      140110194905092622      山西省太原市晋源区
安俊驰      MALE      11011319941109080X      北京顺义区
李珍        FEMALE    140221201212223340      山西省大同市阳高县
东方鑫磊    MALE      330110195907155252      浙江省杭州市余杭区
伏熠彤      MALE      130102200702218535      河北省石家庄市长安区
康楷瑞      MALE      330226201602136039      浙江省宁波市宁海县
云皎洁      FEMALE    140227196307298614      山西省大同市大同县
屈隽雅      FEMALE    330212199802132122      浙江省宁波市鄞州区
尉迟伟祺    MALE      320104201106210695      江苏省南京市秦淮区
伍君洁      FEMALE    130227200003137982      河北省唐山市迁西县
柏越泽      MALE      330282198310152439      浙江省宁波市慈溪市
伏静淑      FEMALE    110109197812141467      北京门头沟区
邵珺琪      FEMALE    120112197410074943      天津津南区
冯家欣      FEMALE    330185198708014451      浙江省杭州市临安市
乐瑶        FEMALE    130102200710264566      河北省石家庄市长安区
皮晴        FEMALE    140110197504193856      山西省太原市晋源区
夏侯远航    MALE      210124196110240199      辽宁省沈阳市法库县
戴熠彤      MALE      140108196707157496      山西省太原市尖草坪区
余珺琪      FEMALE    110107200809262067      北京石景山区
淳于嘉悦    FEMALE    110116196806053486      北京怀柔区
公孙嘉丽    FEMALE    120112195710044579      天津津南区
熊伟诚      MALE      330211200006062230      浙江省宁波市镇海区
慕容立辉    MALE      140121195511107012      山西省太原市清徐县
时筠心      FEMALE    120118197012036616      天津静海区
汪黎昕      MALE      120111196608187264      天津西青区
毕峻熙      MALE      320117197408126889      江苏省南京市溧水区
```

第 2 章　排序和查找

在实际工程中，排序和查找是使用最频繁的基础算法。本章并不讨论各种排序算法的理论基础，只探讨其在工程中的实际应用。

2.1　桶排序（bucket sort）

桶排序的基本思想就是将待排序的数据分别放入不同的桶中，而这些桶之间其实是有序的。根据具体需要，还有可能将所有桶内的数据再进行排序，从而完成整个序列的排序。

常用桶排序的情形包括：按年龄排序、按工龄排序、按成绩排序等。

例如，给定一个年级 *n* 名学生的 C 语言成绩，要求按成绩降序输出每个学生的成绩或者每个成绩段内有多少学生。如果要输出每个成绩有多少学生，那么可以定义一个一维数组用于累加每个成绩的人数，例如，int scoreNum[101]={0}，这个 scoreNum 数组就相当于 101 个桶，表示成绩从 0 到 100 的学生人数。如果要输出每个成绩段内有多少学生，例如，将成绩分成 6 段，A(90~100)，B(80~89)，C(70~79)，D(60~69)，E(50~59)，F(0~49)，则可以定义数组 int rangeNum[6]={0}，这个 rangeNum 数组就相当于 6 个桶，表示每个成绩段的学生人数。

需要指出的是，桶的优化思想不仅仅局限于排序问题。任何一种可以抽象成桶的数据结构，都可以用来作为最简单的算法优化策略。例如，在二维空间中，要查询一个点到另外一个海量数据点的最近点。利用桶排序思想，可以预先将二维平面空间均匀划分成一系列的网格(Grid)，网格中标记其中包含的若干云点。查询最近点时，可以通过计算网格索引号的方法，快速获得若干可能的候选最近点，从而大大对加快最近点的搜索。

2.2　qsort 排序

在 C/C++语言或者数据结构课程教学时，冒泡排序或者选择排序往往是排序章节中首先讲授的内容。因为算法较为简单，所以很多新手碰到排序问题时，往往首先选择冒泡或者选择排序。碰到排序问题，大多数情况下不用思索，直接使用快速排序。确切地说，直接使用 C/C++库函数提供的快速排序函数 qsort(q 是 quick 的简写)，需要引入头文件<stdlib.h>（注：

C++中请尽量使用<cstdlib>)。

C/C++中库函数 qsort 函数声明如下：

```
void qsort (void* base, size_t num, size_t size, int (*compar)(const void*,const void*));
```

第一个参数 base 是待排序的元素数组起始地址；第二个参数 num 是待排序的数组中元素个数；第三个参数 size 是每个元素的大小（字节数），最后一个参数是排序的比较函数。

其中 void*表示空指针，因为 qsort 本身并不知道待排序的元素是什么类型，所以用 void*表示是最合理的。另外，任何一个类型的指针，都能够安全地转换成空指针。size_t 在第 1 章已经介绍，实际就是无符号整型，等价于 unsigned int。int(*compare)(const void*，const void*)是一个函数指针，这里实际上要求传入一个可以回调的比较函数，也就是必须自己定义比较函数。比较函数通常写成如下形式，代码如下：

```
int compareMyType (const void * a, const void * b)
{
  if ( *(MyType*)a <   *(MyType*)b ) return -1;
  if ( *(MyType*)a >   *(MyType*)b ) return 1;
  if ( *(MyType*)a == *(MyType*)b ) return 0;
}
```

比较函数返回值意义见表 2-1。

表 2-1　比较函数返回值意义

值	意义
<0	a 指向的元素要排在 b 指向的元素的前面
>0	a 指向的元素要排在 b 指向的元素的后面
=0	a 指向的元素和 b 指向的元素相等

2.2.1　整型数组的 qsort

掌握 qsort 用法的基本要点，现在开始使用 qsort 排序函数对一个整型数组升序排序，代码如下：

```
#include <stdio.h>          /* printf */
#include <stdlib.h>         /* qsort */
int values[] = { 50, 10, 100, 90, 20, 50, 25 };
int compare (const void * a, const void * b)
{
    if(*(int*)a > *(int*)b)
        return 1;
    else if(*(int*)a < *(int*)b)
        return -1;
    else
        return 0;
}
int main ()
```

```
{
    int i, n;
    n = sizeof(values)/sizeof(int);
    qsort (values, n, sizeof(int), compare);
    for (i=0; i<n; ++i)
        printf ("%d ",values[i]);
    return 0;
}
```

排序结果：

```
10  20  25  50  50  90  100
```

对于数据是整型的元素，上述排序比较函数还可以简化，代码如下：

```
int compare (const void * a, const void * b)
{
    return ( *(int*)a - *(int*)b );
}
```

快速排序最坏时间复杂度是 $O(n^2)$，平均时间复杂度是 $O(n*\lg(n))$。在实际程序中，快速排序速度远远快于冒泡和选择排序。

2.2.2 浮点型数组的 qsort

浮点类型数组升序排序示例，代码如下：

```
#include <stdio.h>          /* printf */
#include <stdlib.h>         /* qsort */
double values[] = { 40.1, 10.8, 100.9, 90.7, 20.3, 90.7, 25.6 };
int compare(const void * a, const void * b)
{
    if (*(double*)a > *(double*)b)        return 1;
    else if (*(double*)a < *(double*)b)     return -1;
    else     return 0;
}
int main ()
{
    int i, n;
    n = sizeof(values)/sizeof(double);
    qsort (values, n, sizeof(double), compare);
    for (i=0; i<n; ++i)
        printf ("%g ",values[i]);
    return 0;
}
```

结果：

```
10.8 20.3 25.6 40.1 90.7 90.7 100.9
```

如果要降序排列，则可以将比较函数修改成以下代码：

```
int compare(const void * a, const void * b)
{
  if (*(double*)b > *(double*)a)       return 1;
  else if (*(double*)b < *(double*)a)     return -1;
  else     return 0;
}
```

对于浮点类型数组，可以将比较函数简化成如下代码吗？代码如下：

```
int compare(const void * a, const void * b)
{
    return *(double*)a - *(double*)b;
}
```

答案是不行！可以想象当 a 指向的浮点数为 10.2， b 指向的浮点数为 10.0， 那么 (int)(10.2−10.0) = (int)(0.2)=0，比较函数居然判断两值相同，这当然是不可接受的！

2.2.3 字符型数组的 qsort

qsort 对字符型数组的升序排序，代码如下：

```
#include <stdio.h>          /* printf */
#include <stdlib.h>         /* qsort */
char values[] = { 'g', 't', 'e', 'w', 'a', 'p' };
int compare (const void * a, const void * b)
{   return *(char*)a -*(char*)b ; }
int main ()
{
  int i, n;
  n = sizeof(values)/sizeof(char);
  qsort (values, n, sizeof(char), compare);
  for (i=0; i<n; ++i)
      printf ("%c ",values[i]);
  return 0;
}
```

结果：

```
a e g p t w
```

修改比较函数为降序排列，代码如下：

```
int compare (const void * a, const void * b)
{   return *(char*)b -*(char*)a ; }
```

这里字符型数组的比较函数也能写成简化形式。

2.2.4 字符串数组的 qsort

字符串数组比较函数的强制类型转化写法要特别注意，代码如下：

```
#include <stdio.h>
#include <stdlib.h>    /*qsort*/
#include <string.h> /*strcmp */

//字典升序比较函数
int StringCmp ( const void *a, const void *b )
{
    return strcmp(*(char **)a, *(char **)b);
}
//字典降序比较函数
int StringCmpGreat ( const void *a, const void *b )
{
    return strcmp(*(char **)b, *(char **)a);
}

void SortString(int bDescending)
{
    char *str[] = {"Eric","Allen", "Beck", "David", "Charlie"};
    int i, n;

    n = sizeof(str)/sizeof(char*);
    if(bDescending)
        qsort(str, n, sizeof(char*), StringCmpGreat);
    else
        qsort(str, n, sizeof(char*), StringCmp);
    for(i=0; i<n; ++i)
    {
        printf("%s", str[i]);
        if(i!=n-1)
            printf(" ");
        else
            printf("\n");
    }
}
int main()
{
    SortString(0);
    SortString(1);
}
```

输出：

```
Allen Beck Charlie David Eric
Eric David Charlie Beck Allen
```

这里需要特别强调，比较函数中传入的指针 a 和 b，它的实际类型是字符指针的指针。所以要写成*(char **)a 的强制转换形式，从而最终将其转换成 char*类型。如果有 char*类型，就可以使用库函数的 strcmp 做比较。strcmp 的返回值意义和 StringCmp 一致，所以可以直接将 strcmp 的返回值作为 StringCmp 的返回值。

在上述代码中还有一个问题，这就是元素数组的类型。

```
char *str[] = {"Eric","Allen", "Beck", "David", "Charlie"};
```

str 是一个指针数组，数组中每个元素都是一个字符类型的指针，而每个指针都指向了一个字符串常量。所以，qsort 调用时，第三个参数一定要写成 sizeof(char*)。请不要用 4 代替 sizeof(char*)，因为在 64 位程序中，指针是 8 字节而不是 4 字节。

如果将字符指针数组换成字符类型的二维数组，代码如下所示：

```
char str[][10] = {"Eric","Allen", "Beck", "David", "Charlie"};
```

那 qsort 及比较函数该如何编写呢？正确方法如下：

```
#include <stdio.h>
#include <stdlib.h>    /*qsort*/
#include <string.h> /*strcmp */

//字典升序比较函数
int StringCmp ( const void *a, const void *b )
{
    return strcmp((char *)a, (char *)b);
}
//字典降序比较函数
int StringCmpGreat ( const void *a, const void *b )
{
    return strcmp((char *)b, (char *)a);
}

void SortString(int bDesending)
{
    char str[][10] = {"Eric","Allen", "Beck", "David", "Charlie"};
    int i, n;

    n = sizeof(str)/(sizeof(char)*10);
    if(bDesending)
        qsort(str, n, sizeof(char)*10, StringCmpGreat);
    else
        qsort(str, n, sizeof(char)*10, StringCmp);
    for(i=0; i<n; ++i)
    {
```

```
            printf("%s", str[i]);
            if(i!=n-1)
                printf(" ");
            else
                printf("\n");
        }
    }
int main()
{
    SortString(0);
    SortString(1);
}
```

因为是二维数组，所以要比较的是以每 10 字节为单位的字符数组，所以 qsort 的第三个参数应该严格地写成 sizeof(char)*10。而比较函数的参数 a 和 b，实际上就是 char*类型，实际开发时这点往往最容易出错。

2.2.5　结构类型数组的 qsort

讨论完基本数据类型的 qsort 用法，然后学习自定义结构类型数组的 qsort 用法。先定义如下学生结构类型：


```
typedef struct _Student
{
    int ID;
    char name[12];
    char city[12];
    float aveScore;
}Student;
```

如果按名字升序排列，排序比较函数应该写成：

```
int StudentNameCmp( const void *a ,const void *b)
{ return strcmp(((Student *)a)->name, ((Student *)b)->name);}
```

如果按学号升序排列，排序比较函数应该写成：

```
//假设两个学号相减不会溢出 int 范围
int StudentIDCmp( const void *a ,const void *b)
{ return (int)(((Student *)a)->ID- ((Student *)b)->ID);}
```

如果按成绩升序排列，排序比较函数应该写成：

```
int StudentScoreCmp( const void *a ,const void *b)
{
  float a2 = ((Student *)a)->aveScore;
  float b2 = ((Student *)b)->aveScore;
  if (a2 > b2) return 1;
```

```
    else if (a2 < b2) return -1;
    else    return 0;
}
```

如果按城市名称升序排列，在同样城市下，按学号升序排列，则排序比较函数应该写成：

```
int StudentCityIDCmp( const void *a ,const void *b)
{
      int ret= strcmp(((Student *)a)->city, ((Student *)b)->city);
      if(ret!=0)
            return ret;
      return (int)(((Student *)a)->ID- ((Student *)b)->ID);
}
```

综合示例代码如下：

```
#include <stdio.h>
#include <stdlib.h>    /*qsort*/
#include <string.h> /*strcmp */

typedef struct _Student
{
      int ID;
      char name[12];
      char city[12];
      float aveScore;
}Student;

int StudentNameCmp( const void *a ,const void *b)
{ return strcmp(((Student *)a)->name, ((Student *)b)->name);}

int StudentIDCmp( const void *a ,const void *b)
{ return ((Student *)a)->ID- ((Student *)b)->ID;}

int StudentScoreCmp( const void *a ,const void *b)
{
     float a2 = ((Student *)a)->aveScore;
     float b2 = ((Student *)b)->aveScore;
   if (a2 > b2) return 1;
   else if (a2 < b2) return -1;
   else    return 0;
}
int StudentCityIDCmp( const void *a ,const void *b)
{
      int ret= strcmp(((Student *)a)->city, ((Student *)b)->city);
      if(ret!=0)
            return ret;
```

```
        return (int)(((Student *)a)->ID- ((Student *)b)->ID);
    }

    void Print(Student *students, int n)
    {
        int i;
        for(i=0; i<n; ++i)
        {
            printf("%-10d%-20s%-20s%-5.2f\n", students[i].ID, students[i].name, students[i].city,students[i].
aveScore);
        }
        printf("\n");
    }

    int main()
    {
        int n;
        Student students[] = {{181101, "Eric", "Nanjing",3.8},
        {181103, "Allen", "Nanjing",3.4},
        {181108, "Eric", "Beijing",3.7},
        {181107, "Brody", "Hangzhou",3.9},
        {181106, "Steward", "Hangzhou",3.8}
         };
        n = sizeof(students)/sizeof(Student);
        printf("Sort by name\n");
        qsort(students, n, sizeof(Student), StudentNameCmp);
        Print(students, n);

        printf("Sort by ID\n");
        qsort(students, n, sizeof(Student), StudentIDCmp);
        Print(students, n);

        printf("Sort by score\n");
        qsort(students, n, sizeof(Student), StudentScoreCmp);
        Print(students, n);

        printf("Sort by city firstly, and sort by ID when their cities are the same\n");
        qsort(students, n, sizeof(Student), StudentCityIDCmp);
        Print(students, n);
    }
```

结果：

```
Sort by name
181103    Allen               Nanjing             3.40
181107    Brody               Hangzhou            3.90
181108    Eric                Beijing             3.70
```

```
181101      Eric        Nanjing      3.80
181106      Steward     Hangzhou     3.80

Sort by ID
181101      Eric        Nanjing      3.80
181103      Allen       Nanjing      3.40
181106      Steward     Hangzhou     3.80
181107      Brody       Hangzhou     3.90
181108      Eric        Beijing      3.70

Sort by score
181103      Allen       Nanjing      3.40
181108      Eric        Beijing      3.70
181106      Steward     Hangzhou     3.80
181101      Eric        Nanjing      3.80
181107      Brody       Hangzhou     3.90

Sort by city firstly, and sort by ID when their cities are the same
181108      Eric        Beijing      3.70
181106      Steward     Hangzhou     3.80
181107      Brody       Hangzhou     3.90
181101      Eric        Nanjing      3.80
181103      Allen       Nanjing      3.40
```

到这里，qsort 所有细节内容都已经阐述。读者可以到各种在线程序评测平台找排序题试试手。作者推荐PTA平台的乙级题目，网址为https://pintia.cn/problem-sets/994805260223102976/problems。排序题目包含 1015、1019、1030 及 1080。

2.3 std::sort 排序

在 C++中，更推荐使用 STL 库函数中的 std::sort，这是一种较为容易掌握并且速度很快的排序方法。注意，这里使用的描述语句是“速度很快的排序”而不是“快速排序”。这是因为 C++ STL 中并没有明确指明 std::sort 使用的排序算法，这也意味不同的平台在 std::sort 函数实现时可能采用不同的排序算法。虽然大部分平台在 std::sort 内部实现时都使用了快速排序算法，但不能认为 std::sort 一定使用了快速排序。

2.3.1 std::sort 基本用法

std::sort 两个重载函数声明如下，需引入头文件<algorithm>。

```
template <class RandomAccessIterator>
   void sort (RandomAccessIterator first, RandomAccessIterator last);
template <class RandomAccessIterator, class Compare>
```

```
void sort (RandomAccessIterator first, RandomAccessIterator last, Compare comp);
```

第一个重载函数只需传入一个迭代器区间，也就是开始和终止位置的两个随机访问迭代器。而第二个重载函数还需传入比较函数或比较对象。这里再次涉及迭代器区间。在 STL 的所有库函数中，如果涉及迭代器区间，那么这个区间一定是半闭半开区间，也就是[first, last)。

在 qsort 函数中，一定要传入一个比较函数，但是在 std::sort 的第一个重载函数中，并没有比较函数。这是因为 qsort 处理的数据都是无符号类型的指针，在编译时候它无法知道数据元素的确切类型。而 std::sort 采用的是 C++模版方法，当数据传到比较函数时，它的类型在编译时就确定了。所以，对于各种基本数据类型，如字符、整型、浮点，都可以直接使用 std::sort 的第一个重载函数进行升序排列。而对于各种类对象，只要在类中重载了 operator<操作符就可以直接调用第一个 std::sort 进行升序排列。不得不说，C++的模板功能功不可没。

下一个问题，就是如何定义 std::sort 的比较函数，也就是调用第二个重载函数时，如何定义 Compare comp。comp 实际上是一个二元函数，它的参数应该是待排序区间中的两个元素，而不是元素的指针。comp 函数的返回值是 bool 类型，当返回值为 true 时，表示 a 排在 b 的前面。毫无疑问，std::sort 的比较函数相比于 qsort 的比较函数显得非常简捷。qsort 比较函数的返回值包括 0、负整数和正整数，分别表示等于、小于和大于。而 std::sort 比较函数的返回值是一个 bool 类型，当值为 true 时表示第一个传入参数对象排在第二个传入参数对象前面，为 false 表示排在后面。最后强调的是 std::sort 比较函数既可以是一个函数指针，也可以是一个函数对象。示例代码如下：

```
#include <iostream>        // std::cout
#include <algorithm>       // std::sort

//double 类型升序比较函数
bool DoubleAscend (double i,double j)
{
    return i<j;
}
//double 类型降序比较函数
bool DoubleDescend (double i,double j)
{
    return i>j;
}

struct CompareClass
{
    bool operator() (double i,double j)
    {
        return (j<i);
    }
} compareDescendObject; //定义了函数对象做比较函数

int main ()
```

```
{
    int i;
    double data[] = {32.5,71.3,12.1,45.4,26.8,80.9,53.2,32.5};
    int n = sizeof(data)/sizeof(double);
    //使用 double 默认升序
    std::sort (data, data+n);
    for(i=0; i<n; ++i)
        std::cout<<data[i]<<" ";
    std::cout << '\n';

    //对前四个元素按降序排序
    std::sort (data, data+4, DoubleDescend);
    for(i=0; i<n; ++i)
        std::cout<<data[i]<<" ";
    std::cout << '\n';

    //对后四个元素按降序排序
    std::sort (data+4, data+n, DoubleDescend);
    for(i=0; i<n; ++i)
        std::cout<<data[i]<<" ";
    std::cout << '\n';

    //对所有元素按降序排序，这里使用了自定义比较类的一个实例化对象
    std::sort (data, data+n, compareDescendObject);
    for(i=0; i<n; ++i)
        std::cout<<data[i]<<" ";
    std::cout << '\n';

    return 0;
}
```

结果：

```
12.1  26.8  32.5  32.5  45.4  53.2  71.3  80.9
32.5  32.5  26.8  12.1  45.4  53.2  71.3  80.9
32.5  32.5  26.8  12.1  80.9  71.3  53.2  45.4
80.9  71.3  53.2  45.4  32.5  32.5  26.8  12.1
```

2.3.2 std::greater 基本用法

对于基本数据类型，STL 已经准备了降序排序的模板函数 std::greater。对于非基本数据类型，如结构体或类，只要其重载 operator<函数，则可以直接使用 std::greater 进行降序排序。使用 std::greater 时需要引入头文件<functional>。参考代码如下：

```
#include <vector>              //std::vector
```

```
#include <functional>          // std::greater
#include <algorithm>           // std::sort
#include <string>              // std::string

    std::vector<int> dataInts;
    ...
    //整型升序排列
    std::sort(dataInts.begin(), dataInts.end());
    //整型降序排列
    std::sort(dataInts.begin(), dataInts.end(), std::greater<int>());

    std::vector<double> datas;
    ...
    //double 类型升序排列
    std::sort(datas.begin(), datas.end());
    //double 类型降序排列
    std::sort(datas.begin(), datas.end(), std::greater<double>());

    std::vector<std::string> strs;
    ...
    //string 类型升序排列
    std::sort(strs.begin(), strs.end());
//string 类型降序排列
    std::sort(strs.begin(), strs.end(), std::greater<std::string>());
```

与降序 std::greater 相对应的是升序 std::less，如下两行代码是完全相同的操作，代码如下：

```
std::sort(dataInts.begin(), dataInts.end());
std::sort(dataInts.begin(), dataInts.end(), std::less<int>());
```

2.3.3 自定义类型排序

有了上述基础，可以使用 std::sort 对结构体成员进行排序。这里先按名字进行字典排序，如果名字相同，则按成绩升序排序，代码如下：

```
#include <iostream>
#include <iomanip>    /*std::setprecision*/
#include <cstring> /*strcmp */
#include <algorithm> /*std::sort*/
using namespace std;
struct Student
{
    int ID;
    char name[12];
    char city[12];
```

```
        float aveScore;
    };
    void Print(const Student *students, int n)
    {
        int i;
        for(i=0; i<n; ++i) cout<<std::left<<setw(10)<<students[i].ID<<setw(20)<<students[i].name<<setw(20)
<<students[i].city<<setw(5)<<std::fixed<<std::setprecision(2)<<students[i].aveScore<<std::endl;
        cout<<endl;
    }

    bool StudentNameAndScoreCmp( Student a , Student b)
    {
        int ret= strcmp(a.name, b.name);
        if(ret!=0)
            return ret<0;
        return a.aveScore<b.aveScore;
    }

    int main()
    {
        int n;
        Student students[] = {{181101, "Eric", "Nanjing",3.8},
        {181103, "Allen", "Nanjing",3.4},
        {181108, "Eric", "Beijing",3.7},
        {181107, "Brody", "Hangzhou",3.9},
        {181106, "Steward", "Hangzhou",3.8}
         };
        n = sizeof(students)/sizeof(Student);
        cout<<"Sort by name\n";
        std::sort(students, students+n, StudentNameAndScoreCmp);
        Print(students, n);
    }
```

输出：

```
Sort by name
181103    Allen               Nanjing             3.40
181107    Brody               Hangzhou            3.90
181108    Eric                Beijing             3.70
181101    Eric                Nanjing             3.80
181106    Steward             Hangzhou            3.80
```

输出结果如期盼的一样完美，更重要的是上述比较函数显然要比 qsort 比较函数容易理解。

当使用 std::sort 和 qsort 做性能比较时，我们动态分配了十万组随机用户数据，结果发现 std::sort 排序耗时要远远超过 qsort。std::sort 采用的算法远逊于 qsort 的快速排序算法吗？答案不是。问题在于本例中 std::sort 的排序函数。再回顾一下上述比较函数，代码如下：

```
bool StudentNameAndScoreCmp( Student a , Student b)
{
    int ret= strcmp(a.name, b.name);
    if(ret!=0)
        return ret<0;
    return a.aveScore<b.aveScore;
}
```

请注意，当前比较函数 StudentNameAndScoreCmp 的两个参数是值传递 (pass by value)。这也意味每次调用 StudentNameAndScoreCmp 时，都存在两次复制构造和两次析构的调用（传递值会产生临时对象的构造）。所以对于自定义数据类型，其比较函数的参数应该采用引用传递方式，正确方法如下：

```
bool StudentNameAndScoreCmp( const Student& a , const Student& b)
{
    int ret= strcmp(a.name, b.name);
    if(ret!=0)
        return ret<0;
    return a.aveScore<b.aveScore;
}
```

2.4 二分查找算法

二分查找又称折半查找，查找速度快，其要求待查表为已经排序的数组或任何顺序存储结构。一个典型的二分查找程序结构如下：

```
int low=0, high=n-1, mid;          //设置当前查找区间上、下界的初值
while(low<=high)
{
    mid=(high+low)/2;
    if(data[mid]==K)
        return mid;                //查找成功返回
    if(data[mid]<K)
        low=mid+1;                 //继续在[mid+1, high]中查找
    else
        high=mid-1;                //继续在[low, mid-1]中查找
}
```

为了避免溢出问题，有时候还需要将 mid=(high+low)/2; 语句替换成：

```
mid=low+(high-low)/2;
```

C 语言库函数提供了一个二分查找函数 bsearch，需要引入头文件<stdlib.h>，函数声明如下：

```
void* bsearch (const void* key, const void* base, size_t num, size_t size,
                    int (*compar)(const void*,const void*));
```

bsearch 的函数参数列表和 qsort 非常类似，只是它多了第一个参数：一个待查找值的指针。如果在数组中成功查找该值，则返回该值所在空间的地址。如果数组中有多个相同的待查找值，则可能返回其中任意一个值的地址。如果查找失败，返回 null 指针。

bsearch 示例代码如下：

```
#include <stdio.h>          /* printf */
#include <stdlib.h>         /* qsort, bsearch*/
int compareInt (const void * a, const void * b)
{
    return ( *(int*)a - *(int*)b );
}

int main ()
{
    int values[] = { 50, 20, 60, 40, 10, 30 };
    int key;
    int * pItem;

    key = 30;
    qsort (values, 6, sizeof (int), compareInt);
    pItem = (int*) bsearch (&key, values, 6, sizeof (int), compareInt);
    if (pItem!=NULL)
        printf ("%d is in the array. Index is %d.\n",*pItem, pItem-values);
    else
        printf ("%d is not in the array.\n",key);
    key = 35;
    pItem = (int*) bsearch (&key, values, 6, sizeof (int), compareInt);
    if (pItem!=NULL)
        printf ("%d is in the array. Index is %d.\n",*pItem, pItem-values);
    else
        printf ("%d is not in the array.\n",key);
    return 0;
}
```

输出：

```
30 is in the array. Index is 2.
35 is not in the array.
```

bsearch 用于字符数组的二分搜索示例代码如下：

```
#include <stdio.h>          /* printf */
#include <stdlib.h>         /* qsort, bsearch*/
#include <string.h>         /* strcmp */
int CompareStr(const void* a, const void *b)
```

```
{
    char* aa = (char*)a;
    char* bb = (char*)b;
    return strcmp(aa, bb);
}

int main ()
{
    char values[][16] = {"adam","tompson", "deck", "kevin","ellen", "beck","charlie" };
    char key[] = "kevin";
    char * pItem;
    int n;
    n = sizeof(values)/(sizeof(char)*16);
    /* sort elements in array: */
    qsort (values, n, 16, CompareStr);
    /* search for the key: */
    pItem = (char*) bsearch (key, values, n, 16,CompareStr);

    if (pItem!=NULL)
        printf ("%s is in the array. Index is %d\n",pItem, (pItem-values[0])/16);
    else
        printf ("%s is not in the array.\n",key);
    return 0;
}
```

C++ STL 库函数中提供了另外一个二分查找函数 std::binary_search。正如 qsort 和 bsearch 是一对，std::binary_search 和 std::sort 也是一对。std::binary_search 有两个重载函数，声明如下，需引入头文件<algorithm>：

```
template <class ForwardIterator, class T>
bool binary_search (ForwardIterator first, ForwardIterator last, const T& val);
template <class ForwardIterator, class T, class Compare>
bool binary_search (ForwardIterator first, ForwardIterator last, const T& val, Compare comp);
```

std::binary_search 示例代码如下：

```
#include <iostream>         // std::cout
#include <algorithm>        // std::binary_search, std::sort
#include <vector>           // std::vector
bool compareInt (int i, int j)
{
    return i<j;
}
int main ()
{
    int data[] = {12,2,3,4,15,4,13,2,13};
```

```
    int n = sizeof(data)/sizeof(int);
    std::sort (data, data+n);
    if (std::binary_search (data, data+n, 13))
        std::cout << "13 found!\n";
    else
        std::cout << "13 not found.\n";

    std::sort (data, data+n, compareInt);
    if (std::binary_search (data, data+n, 8, compareInt))
        std::cout << "8 found!\n";
    else
        std::cout << "8 not found.\n";
    return 0;
}
```

std::binary_search 返回类型是 bool，它无法获得当前查询值在区间内的位置。如果要查询位置，读者可以搜索 std::lower_bound, std::upper_bound 或者 std::equal_range 的用法。

2.5 std::find 查找

std::find 是 C++ STL 库函数中一个通用的查找函数，这个函数并不要求输入的数据集合是已排序的。std::find 最大的优势在于它的易用性。std::find 声明如下，需引入头文件 <algorithm>：

```
template <class InputIterator, class T>
   InputIterator find (InputIterator first, InputIterator last, const T& val);
```

它会返回 [first,last)区间内第一个和 val 相等的元素的迭代器。如果没有找到该元素，函数返回的是 last 迭代器。

std::find 示例代码如下：

```
#include <iostream>      // std::cout
#include <algorithm>     // std::find
#include <vector>        // std::vector

int main ()
{
    int data[] = { 35, 25, 10, 15, 45 };
    int *p, n;

    n = sizeof(data)/sizeof(int);

    p = std::find (data, data+n, 15);
```

```
    if (p != data+n)
        std::cout << "Element found in data, index is " << p-data << '\n';
    else
        std::cout << "Element not found in data\n";

    std::vector<int> dataVec (data,data+n);
    std::vector<int>::iterator it;

    it = find (dataVec.begin(), dataVec.end(), 25);
    if (it != dataVec.end())
        std::cout << "Element found in dataVec, index is " << it-dataVec.begin() << '\n';
    else
        std::cout << "Element not found in dataVec\n";

    return 0;
}
```

输出：

```
Element found in data, index is 3
Element found in dataVec, index is 1
```


可以看出，使用 std::find 的最大优点就是查找的代码非常简捷，不需要写 for 循环语句。读者在实际工作中，可以大胆使用。

2.6 综合编程实例

1. 员工 KPI 排名

从某公司的人事数据库导出员工数据 staff.dat，它的文件格式如下：

```
6
张明    10140155    21
李四    10140156    38
王五    10140157    35
赵六    10140158    15
钱七    10140159    15
王五    10140160    25
```

其中，第一行表示有多少员工，每行存储的信息格式为：

姓名　工号　KPI（绩效考核）

可以看到该数据默认是按工号排序的。

要求：读入该文件后，按 KPI（绩效考核）降序排列，KPI 相同情况下按工号降序排列，然后输出到文件 staffsort.dat 中。

6
李四 10140156 38
王五 10140157 35
王五 10140160 25
张明 10140155 21
钱七 10140159 15
赵六 10140158 15

员工 KPI 排名的综合实例涉及的编程要点有：结构体（或类）定义、文件读入、内存动态分配与释放、数据排序及文件输出。

C 语言版本参考代码如下：

```
#include <stdio.h>
#include <stdlib.h>
#include <string.h>

//定义员工数据结构
typedef struct Employee_
{
    char name[20];
    long ID;
    int KPI;
}Employee;

//从 staff.dat 中读取员工数据
int Read(Employee **staff, int* numStaff)
{
    FILE *file;
    int n, i, KPI;
    char name[20];
    long ID;

    file = fopen("staff.dat", "r");
    if(!file)
    {
        perror("staff.dat can't be opened!");
        return 0;
    }
    fscanf(file,"%d", &n);
    *numStaff = n;
    if((*numStaff)<=0)
    {
        fclose(file);
```

```
        return 0;
    }
    *staff  = (Employee*)malloc(sizeof(Employee)*(*numStaff));
    for(i=0; i<(*numStaff); ++i)
    {
        fscanf(file, "%s %ld %d", name, &ID, &KPI);
        (*staff)[i].ID = ID;
        (*staff)[i].KPI = KPI;
        strcpy((*staff)[i].name, name);
    }
    fclose(file);
    return 1;
}

//排序比较函数
int CompareData(const void*a,const void* b)
{
    Employee *aa = (Employee*)a;
    Employee *bb = (Employee*)b;
    if(aa->KPI<bb->KPI)
        return 1;
    else if(aa->KPI>bb->KPI)
        return -1;
    else
        return bb->ID-aa->ID;
}

//排序函数
void Sort(Employee *staff, int numStaff)
{
    qsort (staff, numStaff, sizeof(Employee), CompareData);
}

//将排序后的员工数据写入 staffsort.dat 文件
void Write(Employee *staff, int numStaff)
{
    int i;
    FILE *file;
    file = fopen("staffsort.dat", "w");
    if(!file)
    {
        perror("staffsort.dat can't be opened to write!");
        return;
    }
```

```
    fprintf(file, "%d\n", numStaff);
    for(i=0; i<numStaff; ++i)
    {
        fprintf(file, "%s %ld %d\n", staff[i].name, staff[i].ID, staff[i].KPI);
    }
    fclose(file);
}

int main()
{
    Employee *staff;
    int numStaff;
    if(!Read(&staff, &numStaff))
        return 1;
    Sort(staff, numStaff);
    Write(staff, numStaff);
    free(staff);
    return 0;
}
```

C++语言版本参考代码如下：


```
#include <iostream>
#include <algorithm>
#include <fstream>

using namespace std;

struct Employee
{
    char name[20];
    long ID;
    int KPI;
};

bool Read(Employee *&staff, int &numStaff)
{
    std::ifstream file;
    int n, i;

    file.open("staff.dat");
    if(file.fail())
    {
        cerr<<"staff.dat can't be opened!";
```

```
            return false;
        }
        file>>n;
        numStaff = n;
        if(numStaff<=0)
        {
            file.close();
            return false;
        }
        staff    = new Employee[numStaff];
        for(i=0; i<numStaff; ++i)
        {
            file>>staff[i].name>>staff[i].ID>>staff[i].KPI;
        }
        file.close();
        return true;
}

bool CompareData(const Employee& a,const Employee& b)
{
        if(a.KPI<b.KPI)
            return false;
        else if(a.KPI>b.KPI)
            return true;
        else
            return a.ID>b.ID;
}

void Sort(Employee *staff, int numStaff)
{
        std::sort(staff, staff+numStaff, CompareData);
}

void Write(Employee *staff, int numStaff)
{
        int i;
        std::ofstream file;
        file.open("staffsort.dat");
        if(!file)
        {
            perror("staffsort.dat can't be opened to write!");
            return;
        }
        file<<numStaff<<"\n";
        for(i=0; i<numStaff; ++i)
```

```
            file<<staff[i].name<<" "<<staff[i].ID<<" "<<staff[i].KPI<<"\n";
        file.close();
    }

    int main()
    {
        Employee *staff;
        int numStaff;
        if(!Read(staff, numStaff))
            return 1;
        Sort(staff, numStaff);
        Write(staff, numStaff);
        delete[] staff;
        return 0;
    }
```

在实际工程中，当不确定处理的数据量的大小时，不要贸然使用固定长度的数组作为数据存储空间。有两点原因：其一，太大的数组处理太小的数据时，必然存在内存浪费现象；其二，小数组处理大数据，必然存在溢出现象。所以，脱离理想主义，最稳妥的方法就是用动态数组或者使用 C++中 STL 容器存储数据。也许有人会反驳说，使用如下方法分配数组不是更好吗？

```
        int n;
        cin>>n;
        int data[n];
```

对于 C/C++十分熟悉的程序员来说，这根本没法编译。但实际上，这是 C99 引入的标准之一，称为 variable-length array，翻译成变量长度的数组。gcc 和 clang 新版本能编译通过，但糟糕的是，广泛使用的 Visual Studio 2015 依旧会编译报错。另外，即使能编译通过，data 的内存在 gcc 中也是被分配在栈空间里。这意味变量长度数组不能太大，否则会溢出栈内存空间，最终导致程序崩溃。

2. MOOC 期终成绩

MOOC 期终成绩这道题出自 PAT 乙级题 1080（作者：陈越），在线测试地址为 https://pintia.cn/problem-sets/994805260223102976/problems/994805261493977088。题目内容如下：

对于在中国大学 MOOC（http://www.icourse163.org/）平台学习“数据结构”课程的学生，想要获得一张合格证书，必须首先获得不少于 200 分的在线编程作业分，然后总评获得不少于 60 分（满分 100）。总评成绩的计算公式为 G =（G 期中×40% + G 期末×60%），如果 G 期中＞G 期末；否则，总评 G 就是 G 期末。这里 G 期中和 G 期末分别为学生的期中和期末成绩。

问题每次考试都产生一张独立的成绩单。请编写程序，把不同的成绩单合为一张。

输入格式：

在第一行输入 3 个整数，分别是 P（做了在线编程作业的学生数）、M（参加期中考试的学生数）、N（参加期末考试的学生数）。每个数都不超过 10000。

接下来有三块输入。第一块包含 P 个在线编程成绩 G 编程；第二块包含 M 个期中考试成绩G期中；第三块包含N个期末考试成绩G期末。每个成绩占一行，格式为：学生学号 分数。其中学生学号为不超过20个字符的英文字母和数字；分数是非负整数（编程总分最高为900分，期中和期末的最高分为100分）。

输出格式：

打印获得合格证书的学生名单。每个学生占一行，格式为：

学生学号　G编程　G期中　G期末　G

如果有的成绩不存在（例如，某人没参加期中考试），则在相应的位置输出“-1”。输出顺序为按照总评分数（四舍五入精确到整数）递减。若有并列，则按学号递增。题目保证学号没有重复，且至少存在1个合格的学生。

输入样例：

```
6 6 7
01234    880
a1903    199
ydjh2    200
wehu8    300
dx86w    220
missing  400
ydhfu77  99
wehu8    55
ydjh2    98
dx86w    88
a1903    86
01234    39
ydhfu77  88
a1903    66
01234    58
wehu8    84
ydjh2    82
missing  99
dx86w    81
```

输出样例：

```
missing  400  -1  99  99
ydjh2    200  98  82  88
dx86w    220  88  81  84
wehu8    300  55  84  84
```

既然有三张成绩单，那么自然可以用三个数组存储这三张成绩单。问题是，从一张表里取一个名字，如何快速地在另外两张表里找到该生对应的两个成绩。如果三张表都没有排序，那么这样查找的时间复杂度是$O(2*n^2)$。毫无疑问，应先对三张表按名字进行字典排序，然后以在线编程作业表中的名字为基准，一个一个比对名字，最后按照规则要求计算最终成绩。这里还需要将成绩合格的学生存储在另一张合格学生表里，当三张输入表都处理完毕后，我们还需要对合格学生表再进行一次排序。综上所述，一共需要四次排序。当然，这里一定使用快速排序。

使用 C++语言的参考代码如下：

```
#include <iostream>
#include <algorithm>
using namespace std;

struct Student
{
      std::string name;               //名字
      short score;                    //成绩
};
struct StudentAdv
{
      std::string name;               //名字
      short p;                        //平时成绩
      short m;                        //期中成绩
      short n;                        //期末成绩
      short ave;                      //综合（最终）成绩
};
//按名字进行比较
bool CompareName(const Student& a, const Student& b)
{
      return a.name<b.name;
}
//按最终成绩递减排序，如果成绩相同则按名字的字典升序排序
bool CompareAdv(const StudentAdv& a, const StudentAdv& b)
{
      if(a.ave>b.ave)
      {
            return true;
      }
      else if(a.ave<b.ave)
      {
            return false;
      }
      else if(a.name<b.name)
            return true;
      else
            return false;
}
#define MAX_N 10001
int main()
```

```
{
    int P, M, N, i, j, k, t, ave;
    Student Ps[MAX_N];  //平时成绩表
    Student Ms[MAX_N]; //期中成绩表
    Student Ns[MAX_N]; //期末成绩表
    StudentAdv results[MAX_N]; //最终合格学生成绩表
    cin>>P>>M>>N;
    //读入平时成绩表
    for(i=0; i<P; ++i)
        cin>>Ps[i].name>>Ps[i].score;
    //读入期中成绩表
    for(i=0; i<M; ++i)
        cin>>Ms[i].name>>Ms[i].score;
    //读入期末成绩表
    for(i=0; i<N; ++i)
        cin>>Ns[i].name>>Ns[i].score;
    //平时成绩按名字排序
    std::sort(Ps, Ps+P, CompareName);
    //期中成绩按名字排序
    std::sort(Ms, Ms+M, CompareName);
    //期末成绩按名字排序
    std::sort(Ns, Ns+N, CompareName);

    //i, j, k 是三张表的下标，  遍历三张表，i, j, k 独立步进；t 是合格学生表的下标
    for(i=0, j=0, k=0, t=0; i<P&&j<M&&k<N; )
    {
        //如果平时成绩小于 200，肯定不合格
        if(Ps[i].score<200)
        {
            ++i;
            continue;
        }
        //如果三张表中当前索引对应的学生名字相同
        if(Ps[i].name==Ms[j].name && Ps[i].name==Ns[k].name)
        {
            //如果期中成绩小于期末成绩，则综合成绩就是期末成绩
            if(Ms[j].score<Ns[k].score)
                ave = Ns[k].score;
            else
                ave = int (Ms[j].score*0.4 + Ns[k].score*0.6+0.5);
            //如果综合成绩不及格，则移动三张表的下标
            if(ave<60)
            {
```

```
            ++i;
            ++j;
            ++k;
            continue;
        }
        //该生合格，记录该生的所有成绩
        results[t].ave = ave;
        results[t].name = Ps[i].name;
        results[t].p = Ps[i].score;
        results[t].m = Ms[j].score;
        results[t].n = Ns[k].score;
        ++t;
        ++i;
        ++j;
        ++k;
    }
    else if(Ps[i].name==Ns[k].name)   //只有平时表和期末表名字相同
    {
        if(Ms[j].name<Ps[i].name)    //如果期中表名字在字典序之前
        {
            ++j;                          //步进期中表中的索引
            continue;
        }
        else
        {
            ave = Ns[k].score;
            if(ave<60)
            {
                ++i;
                ++k;
                continue;
            }
            results[t].ave = ave;
            results[t].name = Ps[i].name;
            results[t].p = Ps[i].score;
            results[t].m = -1;   //未参加期中考试
            results[t].n = Ns[k].score;
            ++t;
            ++i;
            ++k;
        }
    }
    else if(Ps[i].name<Ns[k].name) //平时表中名字小于期末表中的名字
    {
```

```
            ++i;
        }
        else
        {
            ++k;
        }
    }
    //如果期中成绩表已经遍历完成，但是平时成绩表和期末成绩表还没有遍历完
     for(; i<P&&k<N; )
    {
        if(Ps[i].score<200)
        {
            ++i;
            continue;
        }
        if(Ps[i].name==Ns[k].name)
        {
            ave = Ns[k].score;
            if(ave<60)
            {
                ++i;
                ++k;
                continue;
            }
            results[t].ave = ave;
            results[t].name = Ps[i].name;
            results[t].p = Ps[i].score;
            results[t].m = -1;
            results[t].n = Ns[k].score;
            ++t;
            ++i;
            ++k;
        }
        else if(Ps[i].name<Ns[k].name)
        {
            ++i;
        }
        else
        {
            ++k;
        }
    }

    //对合格成绩学生还要再进行一次排序：成绩降序+名字升序
```

```
        std::sort(results, results+t, CompareAdv);
        for(i=0; i<t; ++i)
        {
            cout<<results[i].name<<" "<<results[i].p<<" "<<results[i].m<<" "<<results[i].n<<" "<<results[i].
ave<<endl;
        }
        return 0;
    }
```

第3章　栈、链表与队列

栈、链表和队列是数据结构中重要的三项内容。本章讨论这些数据结构在实际算法中的应用。

3.1　栈

栈(stack)是一种先进后出(FILO: first in, last out)或后进先出(LIFO: last in, first out)的数据结构。其仅允许在一端进行插入和删除操作。这一端称为栈顶，而把相对的另一端称为栈底。向一个栈插入新元素称作进栈或压栈，它是把新元素放到栈顶元素的上面，使之成为新的栈顶元素；从栈删除元素又称作出栈或退栈，它是把栈顶元素弹出，使其相邻的元素成为新的栈顶。栈运行示意图如图 3-1 所示。

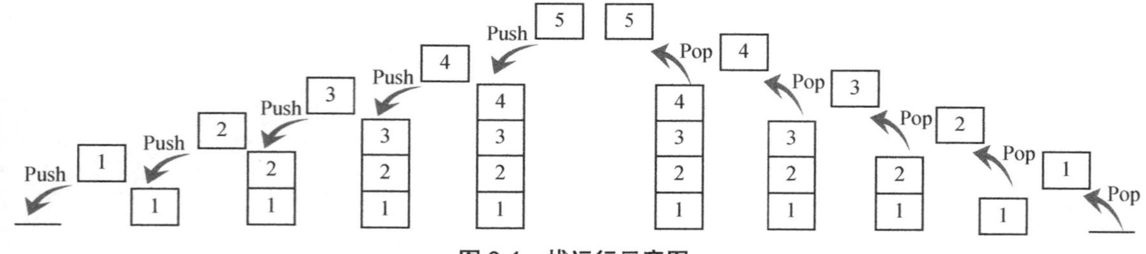

图 3-1　栈运行示意图

栈常见的实现方法是数组，另一种方法是通过单向链表来实现。在 C++中，可以直接使用 STL 的 std::vector, std::list, std::deque 或 std::stack 来表示栈。需要强调的是，std::stack 完全模拟了栈的实现，但其自身并不是 STL 中的标准容器，它是通过容器适配器实现的一个类。

3.1.1　std::stack 基本用法

STL 中 std::stack 声明如下，可见其并不是一个容器，而是用 deque 容器适配器表示。

```
template <class T, class Container = deque<T> > class stack;
```

std::stack 类成员中的常用函数如下：

- empty：判断栈是否为空

- size：返回栈中元素个数
- top：返回栈顶元素
- push：将一个元素压入栈中
- pop：将栈顶元素弹出

std::stack 用法如下：

```
#include <iostream>        // std::cout
#include <stack>       // std::stack

int main ()
{
    std::stack<double> s;
    double v;

    s.push(3.0);
    s.push(3.1);
    s.push(3.2);
    std::cout<<"Elements num "<<s.size()<<". Current top = "<<s.top()<<std::endl;
    while(!s.empty())
    {
        v = s.top();
        s.pop();
        std::cout<<"Current top = "<<v<<std::endl;
    }
    return 0;
}
```

3.1.2 综合编程实例

1. 验证栈操作

假设字母 I 和 O 分别表示入栈和出栈操作。当前栈为空，给定一个由 I 和 O 构成的序列，如果没有出现空栈弹出或者栈溢出，并且最后状态也是栈空，则称该序列是合法的栈操作序列。对每个序列，如果该序列是合法的栈操作则输出 SUCC，否则输出 FAIL。输入第一行给出两个正整数 m 和 n，分别表示栈序列的个数和栈最大容量。例如，输入：

```
3 7
IIIIIIOIIOOOOOOOO
IIIOOIOOIO
IIIOOIOOI
```

则输出：

```
FAIL
SUCC
FAIL
```

对于这道题，当然可以用一个实际的栈去模拟压栈和出栈操作。程序有三个限制条件：
① 压栈之前，栈容量必须小于 n；
② 出栈之前，栈不能为空；
③ 最后每个序列操作结束后，栈必须为空。
可以利用 std::stack，用 C++语言实现上述功能，代码如下：

```
#include <iostream>        // std::cout
#include <stack>       // std::stack

int main ()
{
    std::string str;
    int i, j, m, n;
    bool bSucc;
    std::cin>>m>>n;
    //过滤第一行的回车键
    std::getline(std::cin, str);

    for(i=0; i<m; ++i)
    {
        std::stack<short> s;
        //得到一行序列
        std::getline(std::cin, str);
        //设置序列初始化为正确序列
        bSucc=true;
        for(j=0; j<str.size(); ++j)
        {
            if(str[j]=='I')
            {
                //如果栈已满，再入栈不满足程序条件
                if(s.size()==n)
                {
                    bSucc=false;
                    break;
                }
                s.push(1);
            }
            else
            {
                //如果栈为空，再出栈不满足程序条件
                if(s.empty())
                {
                    bSucc=false;
                    break;
```

```
                }
                s.pop();
            }
        }
        if(bSucc)
        {
            //如果最后栈不为空，不满足程序条件
            if(!s.empty())
                bSucc=false;
        }
        std::cout<<(bSucc?"SUCC":"FAIL")<<std::endl;
    }
    return 0;
}
```

对于 C 程序，本题可以简单地利用栈元素计数器实现上述功能，代码如下：

```
#include <stdio.h>          // printf, fgets
#define MAX_S 1024
int main ()
{
    int i, j, m, n, size;
    short bSucc;
    char str[MAX_S];
    scanf("%d%d", &m, &n);
    //过滤第一行的回车键
    fgets(str, MAX_S, stdin);
    for(i=0; i<m; ++i)
    {
        size = 0;
        bSucc=1;
        //读取一行输入
        fgets(str, MAX_S, stdin);
        for(j=0; str[j]!='\n'; ++j)
        {
            if(str[j]=='I')
            {
                //如果栈已满，再压入元素则错误
                if(size>=n)
                {
                    bSucc=0;
                    break;
                }
                ++size;
            }
```

```
                else
                {
                    //如果栈已空，再弹出元素错误
                    if(size==0)
                    {
                        bSucc=0;
                        break;
                    }
                    --size;
                }
            }
            //如果栈非空，错误
            if(bSucc && size!=0)
                bSucc=0;
            if(bSucc)
                printf("SUCC\n");
            else
                printf("FAIL\n");
        }
        return 0;
    }
```


2. 汉诺塔非递归求解问题

汉诺塔是个经典的递归问题。有三根杆（编号 A、B、C），在 A 杆上自下而上、由大到小按顺序放置若干盘子。游戏的目标：把 A 杆上的盘子全部移到 C 杆上，并仍保持原有顺序叠好。操作规则：每次只能移动一个盘子，并且在移动过程中三根杆上都始终保持大盘在下，小盘在上，操作过程中盘子可以置于 A、B、C 任一杆上。

为了将这 n 个盘子从 A 杆移动到 C 杆，可以做以下三步工作。

（1）以 C 盘为中介，从 A 杆将 1 至 n-1 号盘移至 B 杆。

（2）将 A 杆中剩下的第 n 号盘移至 C 杆。

（3）以 A 杆为中介，从 B 杆将 1 至 n～1 号盘移至 C 杆。

根据以上递归规则，可以写出如下递归函数，代码如下：

```
void Hanoi(char src, int n, char media, char des)
{
    if(n==1)
    {
        printf("%c->%c\n", src, des);
        return;
    }
    Hanoi(src, n-1, des, media);
    printf("%c->%c\n", src, des);
    Hanoi(media,n-1, src, des);
}
```

理论上，任何一个递归解法都可以采用显示栈的非递归方式进行替换。递归方法实现简单，但是递归深度不能太深，否则会溢出程序栈空间。因为一个程序的栈空间通常也只有几百 KB 至几 MB 的大小，在 Visual Studio 中可以设置程序栈的默认大小。如果要克服递归栈空间的限制问题，就只能使用非递归的显示栈进行替换，因为显示栈的内存可以分配在堆区。

在 C++中，可以利用 std::stack 实现汉诺塔的非递归算法，代码如下：

```
#include <iostream>        // std::cout
#include <stack>       // std::stack

//模拟栈中的数据，也就是对数据打包
struct HanoiData
{
    HanoiData() { }
    HanoiData(char a, int num, char b, char c):src(a), n(num), media(b), des(c)    { }
    char src;          //原杆
    int n;             //当前 src 杆上的盘子个数
    char media;        //中介杆
    char des;          //目标杆
};

int main ()
{
    //高速 I/O，iostream 不与 cstdio 同步
    std::ios::sync_with_stdio(false);
    //数据栈
    std::stack<HanoiData> s;
    HanoiData d;
    int n;
    std::cin>>n;

    //初始栈
    s.push(HanoiData('a', n, 'b', 'c'));
    while(!s.empty())
    {
        //获得栈顶元素
        d = s.top();
        //弹出栈顶元素
        s.pop();
        if(d.n==1)
        {
            std::cout<<d.src<<" -> "<<d.des<<"\n"; //避免用 std::endl
        }
        else
```

```
        {
            //显示栈必须倒过来先处理后续问题
            s.push(HanoiData(d.media, d.n-1, d.src, d.des));
            //不能直接打印，而是要继续通过入栈打印
            s.push(HanoiData(d.src, 1, d.media, d.des));
           // std::cout<<d.src<<"->"<<d.des<<std::endl;
            s.push(HanoiData(d.src, d.n-1, d.des, d.media));
        }
    }
    return 0;
}
```

需要强调的是，显示栈和递归栈在处理问题时，先后顺序是倒置的。

如果要将上述C++代码转换成C代码，读者只需要写出栈的创建、压栈、出栈、栈顶访问、栈元素个数函数即可，这里不再赘述。

3. 火车调度问题

车站调度示意图如图3-2所示。其中列车需从A方向驶入车站，从B方向驶出车站。其中火车进站顺序编号为1～n。为了火车调度需要，可以借助中转站C。C可以停放任意多辆火车，但驶入C的火车必须按照相反的顺序驶出。对于每列火车，一旦它从A移入C就不能再回到A；一旦从C移入B就不能再回到C。也就是任意时刻，只有A→C和C→B。

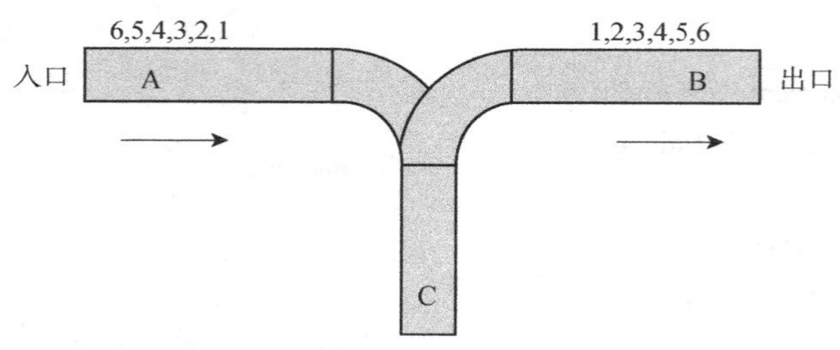

图3-2 车站调度示意图

输入格式，第一行是火车数量n，下一行是从B驶出的火车序列编号，你需要判断该调度是否存在，如果是打印YES，不存在则打印NO。

例如，输入：

```
6
6 5 4 3 2 1
4
1 2 3 4
5
5 4 1 3 2
```

输出：

```
YES
YES
NO
```

毫无疑问 C 作为调度导轨实际上就是一个栈，而 B 导轨上的火车驶出顺序既可以用数组表示，也可以用链表表示。以下是一种利用 std::stack 和 std::list 的 C++解法，代码如下：

```
#include <iostream>
#include <stack>
#include <list>

using namespace std;

int main()
{
    int i, n, id;
    while(cin>>n)
    {
        std::list<int> B;
        std::stack<int> C;
        for(i=0; i<n; ++i)
        {
            cin>>id;
            B.push_back(id);
        }
        id = 1;
        while(!B.empty())
        {
            if(!C.empty() && C.top()==B.front())
            {
                C.pop();
                B.pop_front();
            }
            else
            {
                C.push(id++);
                if(id>n+1)
                    break;
            }
        }
        if(C.empty())
            cout<<"YES"<<std::endl;
        else
            cout<<"NO"<<std::endl;
    }
    return 0;
}
```

如果要将上述程序改成 C 语言版本，则只需要将 stack 和 list 用数组进行模拟即可。这里不再赘述。

4. 多堆栈问题——队列模拟

用两个堆栈A和B模拟出队列Q的入队和出队操作。这是一道经典的堆栈模拟队列问题，也是一些公司的常用面试题。本题首先给出两个正整数 $n1$ 和 $n2$，表示堆栈A和B的最大容量。随后给出一系列的队列操作：P a表示将整数a入列；O表示将队首元素出队；E表示输入结束。输出格式要求对输入中的每个出队（O）操作，打印出队数字，如果当前队列为空则输出错误信息Empty。如果入队时队列已满，则输出Full。

例如，输入：

```
3 4
P 1 P 2 P 3 P 4 P 5 P 6 O P 7 O P 8 O P 9 O O O O O E
```

输出:

```
1
Full
2
Full
3
4
5
6
9
Empty
```


本题解题思路是始终把入队动作放在A栈来操作，而把出队动作放在B栈来操作。所以一旦有出队动作时，就要检测B栈。如果当前B栈为空，则需要将A栈内容全部倒入B栈；如果B栈不为空，则直接从B栈弹出栈顶元素。根据上述规则，两个栈相互作用的队列模拟逻辑如下：

```
if(有元素入列)
   if(A栈元素个数小于n)
        则让该元素入A栈
   else
        if(B栈不为空）
            则队列满，不能入列
        else if(B栈为空)
            将A栈中所有元素先弹入B栈，再将该元素弹入A栈
else   //有元素出列
    if(B栈不为空）
        将B栈顶元素弹出
    else
            if(A栈不为空)
                将A栈中元素全部弹入B栈，然后弹出B栈顶元素
```

else if(A 栈为空)
    队列空，不能出列

此处 n 是 min($n1$, $n2$)。根据以上逻辑，示例代码如下：

```
#include <iostream> //std::cin, std::cout
#include <stack>    //std::stack
using namespace std;

int main()
{
    int n1, n2, n, v;
    char op;
    std::stack<int> A, B;
    cin>>n1>>n2;
    n = std::min(n1, n2);

    while(true)
    {
        cin>>op;
        if(op=='P')
        {
            cin>>v;
            if(A.size()<n)
                A.push(v);
            else
            {
                if(!B.empty())
                    cout<<"Full"<<endl;
                else
                {
                    while(!A.empty())
                    {
                        B.push(A.top());
                        A.pop();
                    }
                    A.push(v);
                }
            }
        }
        else if(op=='O')
        {
            if(!B.empty())
            {
                cout<<B.top()<<endl;
```

```
                    B.pop();
                }
                else
                {
                    if(!A.empty())
                    {
                        while(!A.empty())
                        {
                            B.push(A.top());
                            A.pop();
                        }
                        cout<<B.top()<<endl;
                        B.pop();
                    }
                    else
                        cout<<"Empty"<<std::endl;
                }
            }
            else if(op=='E')
                break;
    }
    return 0;
}
```

3.2 链表

链表（Linked List）是一种非连续、非顺序的存储结构。链表由一系列结点组成，它允许插入和移除表上任意位置上的结点，但是无法随机存取。链表通常有三种表示类型：单向链表，双向链表及循环链表。

单链表（Singly Linked List）又称单向链表，它的结点包含两个部分：存储数据的数据域和存储下一个结点地址的指针域。单链表结点结构通常定义如下：

```
struct Node
{
    ElementType  data;        //数据
    Node* next;               //指向下一个结点的指针
};
```

单链表如图 3-3 所示。

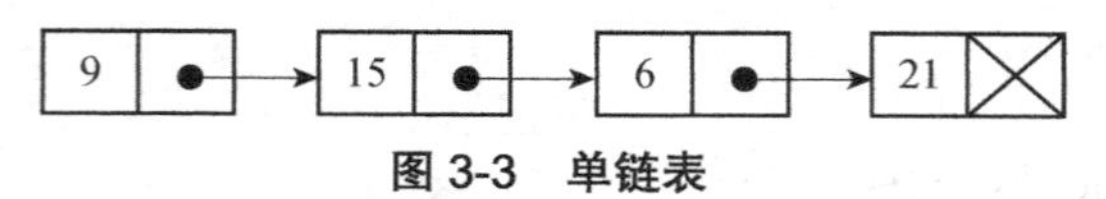

图 3-3　单链表

双向链表(Double Linked List)是单向链表的改进。在双向链表的结点中，除含有数据域和存储下一个结点地址的指针域外，还有一个存储直接前驱结点地址的指针域。双向链表结点结构通常定义如下：

```
struct Node
{
    ElementType data;          //数据
    Node* next;                //指向下一个结点的指针
    Node* pre;                 //指向前一个结点的指针
};
```

双向链表如图 3-4 所示。

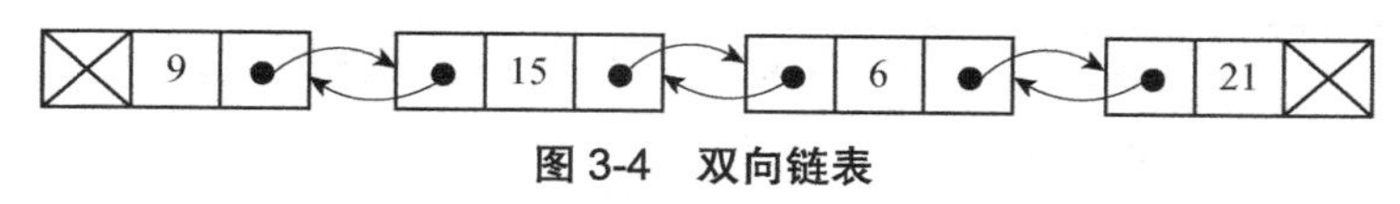

图 3-4　双向链表

循环链表（Circular Linked List）与单链表的结点结构相同，所不同的是，循环链表的最后一个结点的指针是指向该循环链表的第一个结点或者表头结点，从而构成一个环形的链。循环链表如图 3-5 所示。

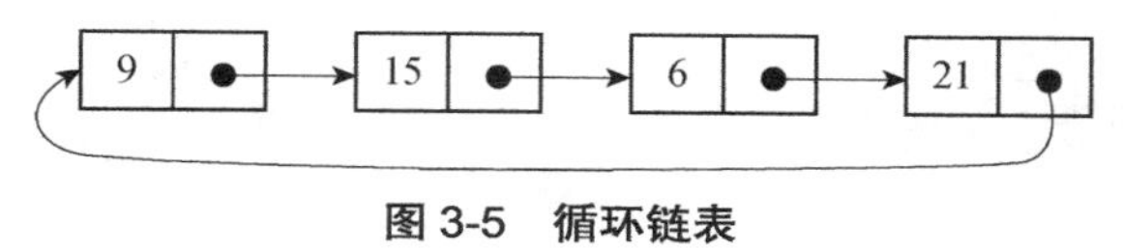

图 3-5　循环链表

3.2.1　std::list 基本用法

STL 中 std::list 是一个顺序容器，它的插入和删除操作都是常量时间。std::list 在内部采用双向链表结构，所以它能两端迭代。在 C++11 标准中，STL 还引入了 std::forward_list，这是一种单向链表结构的容器。std::list 和 std::forward_list 的主要缺点是无法随机访问，即无法通过下标访问；最大优点是插入和删除都是常量时间。std::list 声明如下：

```
template < class T, class Alloc = allocator<T> > class list;
```

std::list 常用成员函数如下：

- front　　访问第一个元素
- back　　访问最后一个元素
- push_front　　在起始位置插入一个元素
- pop_front　　删除第一个元素
- push_back　　在末尾位置添加一个元素
- pop_back　　删除最后一个元素
- insert　　在指定迭代器位置插入一个或多个元素
- erase　　删除迭代器指定位置或区间的一个或多个元素
- empty　　判断当前链表是否为空
- size　　获得当前链表元素个数
- remove　　从链表中删除具有某个值的元素

这里只强调一个成员函数 erase，也就是删除链表结点函数。std::list 的 erase 有两个重载函数：

```
iterator erase (iterator position);
iterator erase (iterator first, iterator last);
```

第一个重载函数是从链表中删除单个元素；第二个重载函数是删除区间[first,last)内的所有元素。erase 的返回值是一个迭代器，这个迭代器指向被删除元素的下一个元素位置。所以，如果当删除一个元素后，希望能继续遍历当前链表时，通常需要写成如下形式：

```
it = mylist.erase(it);
```

std::list 的简单用法如下：

```
#include <iostream>
#include <list>

int main ()
{
    std::list<int> mylist;
    std::list<int>::iterator it;

    // initial values in list
    // 2 4 6 8 10
    for (int i=1; i<=5; ++i)
        mylist.push_back(2*i);

    it = mylist.begin();
    // it points now to number 4
    ++it;

    // 2 15 [4] 6 8 10
    mylist.insert (it, 15);

    // 2 15 30 30 [4] 6 8 10
    mylist.insert (it, 2, 30);

    // it points now to the second 30
    // 2 15 30 [30] 4 6 8 10
    --it;

    //erase the second 30, and it points to the next element 4
    // 2 15 30 [4] 6 8 10
    it = mylist.erase(it);

    for (it=mylist.begin(); it!=mylist.end(); ++it)
        std::cout << ' ' << *it;
```

```
    std::cout << '\n';
    return 0;
}
```

3.2.2 综合编程实例

1. 约瑟夫环问题

已知 n 个人（以编号 1，2，3，…，n 分别表示）围坐在一张圆桌周围。从编号为 k 的人开始报数，数到 m 的那个人出列；他的下一个人又从 1 开始报数，数到 m 的那个人又出列；依此规律重复下去，直到圆桌周围的人全部出列。问最后剩下的一个人的编号是多少？或者依次打印出列的人的编号。

约瑟夫环问题是经典的链表问题。由于人都是围坐在圆桌周围，所以不难想象，解决该问题需要使用环形链表，代码如下：

```
#include <iostream>
#include <list>

int main ()
{
    std::list<int> table;
    int i, n, k, m;
    //such as 6 1 3
    std::cin>>n>>k>>m;
    for(i=1; i<=n; ++i)
        table.push_back(i);
    std::list<int>::iterator it=table.begin();
    for(i=1; i<k; ++i, ++it);
    int count=1;
    while(table.size()>1)
    {
        if(count++==m)
        {
            //打印出列的人编号
            std::cout<<*it<<std::endl;
            //利用 erase 删除，再指向一个元素
            it = table.erase(it);
            count=1;
        }
        else
            ++it;
        //最后，从头循环,所以构成环形链表
        if(it==table.end())
            it=table.begin();
```

```
        }
        std::cout<<"The left person id = "<<table.front()<<".\n";
        return 0;
}
```

2. 链表的倒数第 *n* 项

给定串联的一系列链表中结点的数据值，目标是要查找倒数第 *n* 个结点的数字。首先输入一个正整数 *n*，随后输入若干整数，输出倒数第 *n* 个位置上的数据值，如果这个位置不存在，输出错误信息 ERROR。

例如，输入：

```
5  13  16  2  4  16  32  64  28  25  51  76
```

则输出：

```
64
```

题目既然提到数据都是链表中结点的数据值，所以不难用 std::list 写出如下的基于链表的解法，代码如下：


```
#include <iostream>    //std::cout
#include <list>           //std::list
#include <string>        //std::string
#include <sstream>     //std::istringstream

int main ()
{
        int i, n, size=0;
        std::list<int> data;
        std::string str;
        //得到整个输入行的字符串
        std::getline(std::cin, str);
        //将字符串 str 打包到 istringstream 输入流中
        std::istringstream iss(str);
        //从流中获得 n，即倒数位
        iss>>n;
        //从流中持续获得数据
        while(iss>>i)
        {
    //如果链表中已经有 n 个数据，所以需要删掉第一个结点，并把新结点加到链表尾部
                if(size= =n)
                {
                        data.erase(data.begin());
                        data.push_back(i);
                }
                else
```

```
            {
                ++size;
                data.push_back(i);
            }
        }
        if(size!=n)
            std::cout<<"ERROR"<<std::endl;
        else
            std::cout<<data.front()<<std::endl;
        return 0;
    }
```

以上算法完全正确，但存在的问题是如果链表非常大，则程序中需要频繁地删除第一个结点，并需向链表尾部添加新结点。而删除和添加操作希望把数据的存储变成一个循环链表。所以，不难利用数组和当前存储位置的索引，重写算法，代码如下：

```
#include <iostream> //std::cout
#include <vector>   //std::vector
#include <string>    //std::string
#include <sstream>   //std::istringstream
int main ()
{
    int i, n, realLen=0;
    std::vector<int> data;
    std::string str;
    //得到一行输入
    std::getline(std::cin, str);
    //将输入字符串打包到 istringstream 流中
    std::istringstream iss(str);
    //获得倒数位置 n
    iss>>n;
    //在数组中分配 n 个数据空间
    data.resize(n);
    //从流中持续获得数据
    while(iss>>i)
    {
        //用取模的方法实现循环利用数组
        data[realLen%n] = i;
        ++realLen;
    }
    if(realLen<n)
        std::cout<<"ERROR"<<std::endl;
    else
        std::cout<<data[realLen%n]<<std::endl;
    return 0;
```

```
}
```

3.3 队列

队列（queue）只允许在表前端进行删除操作，而在表的后端进行插入。在队列中插入一个元素称为入队，从队列中删除一个元素称为出队。因为队列只允许在一端插入，在另一端删除，所以最早进入队列的元素最先从队列中删除，也就是先进先出（FIFO，First In First Out）。队列示意图如图 3-6 所示。

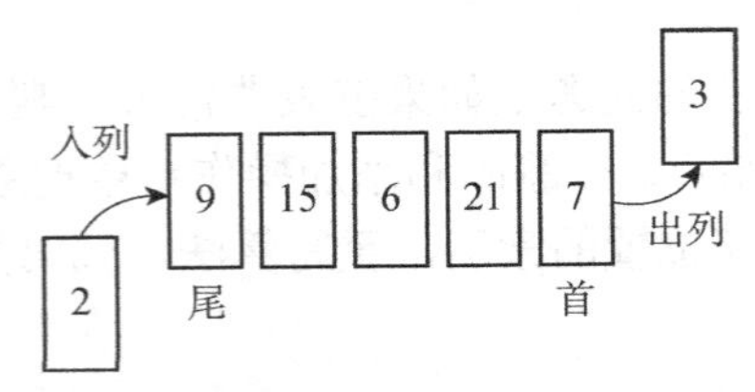

图 3-6 队列示意图

通常使用两种方法实现队列。

① 环形缓冲。

② 双向链表。

所谓环形缓冲，实际上是静态或动态分配的一个数组，它设置两个指针。一个指针 front 指向队头元素，另一个指针 rear 指向下一个入队元素的存储位置也就是队尾。每次在队尾插入一个元素时，rear 增 1；每次在队头删除一个元素时，front 增 1。随着插入和删除操作的进行，队列元素的个数不断变化。一旦 rear 指针增 1 或 front 指针增 1 时超出了所分配的数组空间，就让它指向这片连续空间的起始位置，也就是采用取余运算 rear=(rear+1)%size 和 front=(front+1) %size 来实现环形缓冲存储。这就相当于把队列空间想象成一个环形空间，用这种方法管理的队列也就称为循环队列。 在循环队列中，当队列为空时，front 等于 rear，而当所有队列空间全占满时，也有 front 等于 rear。所以为了区别这两种情况，规定循环队列最多只能有 size-1 个队列元素，当循环队列中只剩下一个空存储单元时，队列就已经满了。因此，队列为空的条件依旧是 front=rear，但队列满的条件修改为 front=（rear+1）%size。对于 C 语言开发者来说，利用这种数组实现的循环队列是必须掌握的基本知识。

相对于环形缓冲，基于双向链表的队列，虽然创建和删除结点效率较低，但可以动态增大队列长度。新元素进来时，被插入到链表的尾部，而读取的时候是从链表的头部开始读取的。每次读取一个元素，释放一个元素。

3.3.1 std::queue 基本用法

std::queue 是 STL 中的一个队列类。和 std::stack 一样，std::queue 实际上并不是 STL 中的标准容器，它只是一个容器适配器，std::queue 声明如下：

```
template <class T, class Container = deque<T> > class queue;
```

可以看出 std::queue 实际上就是从标准容器 std::deque 实例化的类。std::queue 常用成员函数如下：

- empty　　判断队列是否为空
- size　　返回队列元素个数
- front　　得到队首元素
- back　　得到队尾元素
- push　　向队尾添加一个元素
- pop　　删除队首元素

std::queue 用法示例代码如下：

```
#include <iostream>              // std::cin, std::cout
#include <queue>                 // std::queue

int main ()
{
    std::queue<double> q;
    double v;

    std::cout << "请输入一些正实数(输入负数表示结束)到队列:\n";

    while(true)
    {
        std::cin >> v;
        if(v<0)
            break;
        q.push (v);
    }

    std::cout << "依次弹出队首元素:";
    while (!q.empty())
    {
        std::cout << ' ' << q.front();
        q.pop();
    }
    std::cout << '\n';
    return 0;
}
```

3.3.2 std::deque 基本用法

deque 是 double ended queue 的简写，std::deque 也就是 STL 中的双向(又称双端)队列容器。双向队列，意味同时可以在队首和队尾添加或删除元素。使用 std::queue 需要引入头文件 <deque>，std::deque 部分常用类成员函数如下：

- push_back　　添加元素到队列尾部
- push_front　　添加元素到队列首部
- pop_back　　从队列尾部弹出一个元素
- pop_front　　从队列首部弹出一个元素
- insert　　在任意位置插入元素
- erase　　在任意位置删除元素

std::deque 示例代码如下：

```
#include <iostream>             // std::cin, std::cout
#include <deque>                // std::deque

int main ()
{
    std::deque<double> q;
    double v;

    std::cout << "请输入一些正实数(输入负数表示结束)到队列:\n";

    while(true)
    {
        std::cin >> v;
        if(v<0)
            break;
        q.push_back(v);
    }

    std::cout << "依次弹出队首元素:";
    while (!q.empty())
    {
        std::cout << ' ' << q.front();
        q.pop_front();
    }
    std::cout << '\n';
    return 0;
}
```

3.3.3　综合编程实例：卡片游戏

桌上有一摞牌，从第一张牌开始从上往下依次编号为 1～n。当至少还剩两张牌时进行以下操作：把第一张扔掉，然后把新的第一张放到整摞牌的最后。要求依次打印扔掉的牌及最后剩下的牌。

例如，输入:

```
6
```

输出

```
1 3 5 2 6 4
```

不难看出卡片游戏问题完全可以用双向队列进行模拟：扔掉第一张牌，也就是弹出队首元素；把新的第一张牌放到整摞牌最后，就是把新队首元素添加到队尾，然后弹出新队首元素。所以，利用 std::deque 不难写出如下代码：

```
#include <iostream>
#include <deque>
using namespace std;
int main()
{
    int i, n;
    bool bFirst=true;
    std::deque<int> cards;
    cin>>n;
    //初始化一堆牌
    for(i=1; i<=n; ++i)
        cards.push_back(i);
    //当牌数量大于 2
    while(cards.size()>=2)
    {
        if(!bFirst)
            cout<<" ";
        //打印即将扔掉的第一张牌
        cout<<cards.front();
        //扔掉第一张牌
        cards.pop_front();
        //把新的第一张牌（即原来第二张牌）挪到最后一张
        cards.push_back(cards.front());
        cards.pop_front();
        bFirst=false;
    }
    if(cards.size())
    {
        if(!bFirst)
            cout<<" ";
        cout<<cards.front();
    }
    return 0;
}
```

第 4 章　树与图

树和图是许多高级算法的基础，本章讨论这两种数据结构在实际开发中的使用方法。

4.1　二叉树

树是数据结构中一项重要的内容。在树的结构中，必须有一个根结点 r，余下的结点构成 r 的子树。树最重要的一种表示方法是二叉树（Binary Tree）。当二叉树非空时，有一个根结点 r，余下的结点被安排在两个分叉上，分别称为 r 的左子树和右子树。二叉树中任何一个结点最多只有两个子结点。

4.1.1　完全二叉树

二叉树的高度有如下特性：若二叉树的高度为 h，则该二叉树至少有 h 个元素，最多有 2^h-1 个元素；对于包含 n 个元素的二叉树，它的最大高度为 n，最小高度为 $\lceil \log_2(n+1) \rceil$。

当高度为 h 的二叉树恰好有 2^h-1 个元素时，称为满二叉树(full binary tree)。对高度为 h 的满二叉树中的元素按从上到下、从左到右的顺序依次编号为 1，2，…，2^h-1。如果从满二叉树中删除 t 个元素，这 t 个元素的编号为 2^h-i, $1\leqslant i\leqslant t$，这样得到的二叉树称为完全二叉树（Complete Binary Tree）。含有 n 个元素的完全二叉树的深度为 $\lceil \log_2(n+1) \rceil$。

三个完全二叉树如图 4-1 所示。

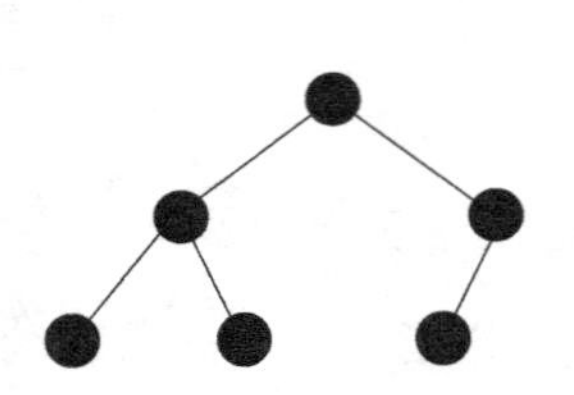
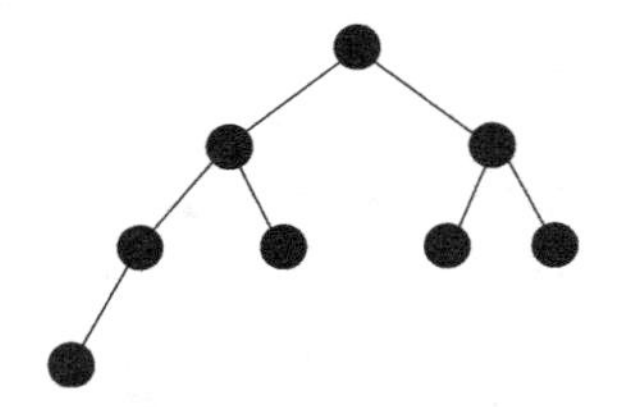
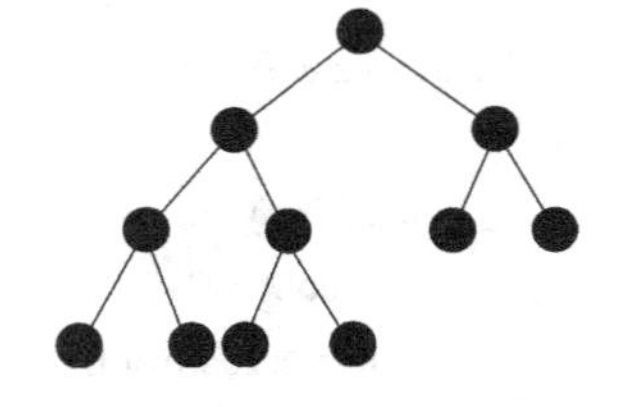

图 4-1　三个完全二叉树

从完全二叉树的定义不难分析其结点编号（$1\leqslant i\leqslant n$）有如下特性。

（1）编号为 1 的结点是该二叉树的根。

（2）除根外，编号为 i 的结点的父结点编号为$\lfloor i/2 \rfloor$，其左孩子的编号为 $2i$，右孩子编号为 $2i+1$。

（3）如果 $2i>n$ 时，则该结点无左子树。如果 $2i+1>n$，该结点无右子树。

4.1.2 二叉树遍历

二叉树的遍历方法有四种。

- 前序遍历 PreOder
- 中序遍历 InOrder
- 后序遍历 PostOrder
- 层次遍历 LevelOrder

其中，前序遍历可以用递归方法描述，代码如下：

```
void PreOrder(TreeNode *t)
{
    if(!t)
        return;
    Visit(t) ;                  // 访问结点
    PreOrder(t->left) ;         // 前序遍历左子树
    PreOrder(t->right) ;        // 前序遍历右子树
}
```

中序遍历可以用递归方法描述，代码如下：

```
void InOrder(TreeNode *t)
{
    if(!t)
        return;
    InOrder(t->left);           //中序遍历左子树
    Visit(t) ;                  // 访问结点
    InOrder(t->right);          //中序遍历右子树
}
```

后序遍历可以用递归方法描述，代码如下：

```
void PostOrder(TreeNode *t)
{
    if(!t)
        return;
    PostOrder(t->left);         //后序遍历左子树
    PostOrder(t->right);        //后序遍历右子树
    Visit(t) ;                  // 访问结点
}
```

二叉树的层次遍历，又称为逐层遍历。在逐层遍历过程中，是按照从顶层到底层的次序访问树结点，在同一层中，再按照从左到右的次序进行访问。所以它是一种宽度优先遍历（Breadth-first Search）方法，需要使用先进先出的队列作为种子结点的存储空间。层次遍历方法描述代码如下：

```
void LevelOrder(TreeNode *t)
```

```
{
    TreeNode* node;
    std::queue<TreeNode*> seeds;
    if(t)
        seeds.push(t);
    while(!seeds.empty())
    {
        //从队首弹出即将访问的结点
        node = seeds.front();
        seeds.pop();
        //访问当前结点
        Visit(node);
        // 将左孩子放入队列
        if (node->left)
            seeds.push(node->left);
        // 将右孩子放入队列
        if (node->right)
            seeds.push(node->right);
    }
}
```


二叉树的宽度优先遍历是关于二叉树的算法中出现频率较高的算法，大家需要掌握。

4.1.3 手写二叉树的遍历

给出一棵二叉树的图片，如何快速地手写二叉树的各种遍历结果是一个有趣的问题。这里提供一种走墙根方法。

首先，想象树中每个结点都是一个圆形塔楼，将树中每条边想象成一定宽度的城墙。走墙根方法，就是从根结点所在的圆形塔楼的西边作为起点，一直沿着墙根走，一直到达根结点所在圆形塔楼的东边为止。当然，每次走墙根的时候，你可能需要在圆形塔楼墙边用粉笔绘制一个记号。根据记号顺序就能够得到每种遍历方法的顺序。

对于前序遍历，记号绘制在每个塔楼的西边。前序遍历：G，C，B，A，E，D，F，I，H，J，如图 4-2 所示。

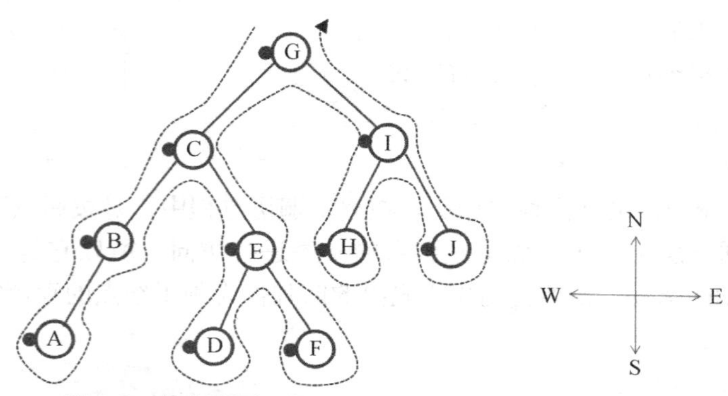

图 4-2　前序遍历：G，C，B，A，E，D，F，I，H，J

对于中序遍历，记号绘制在每个塔楼的南边。中序遍历：A，B，C，D，E，F，G，H，I，J，如图 4-3 所示。

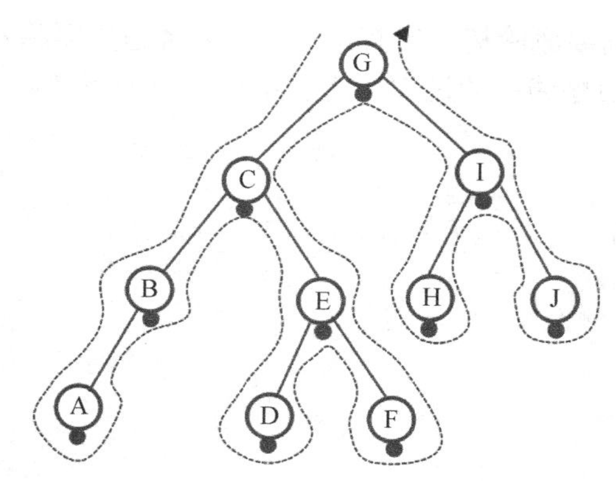

图 4-3　中序遍历：A，B，C，D，E，F，G，H，I，J

对于后序遍历，记号绘制在每个塔楼的东边。后序遍历：A，B，D，F，E，C，H，J，I，G，如图 4-4 所示。

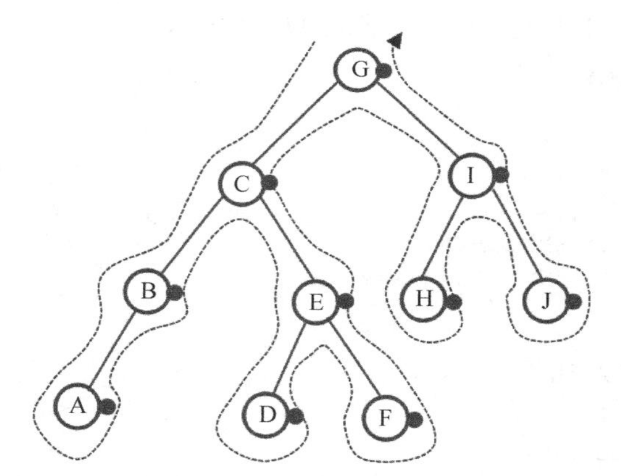

图 4-4　后序遍历：A，B，D，F，E，C，H，J，I，G

4.1.4　二叉树高度计算

不难想象，二叉树中任意一个结点的高度值实际上就是它左右两棵子树中较大一个结点的高度值再加上 1，所以可以递归地写出如下结点高度计算函数，代码如下：

```
int Height(TreeNode *t)
{
    if (!t)
        return 0;
    int leftHeight = Height(t->left);         // 左子树的高度
    int rightHeight = Height(t->right);     // 右子树的高度
    return max(leftHeight, rightHeight)+1;
}
```

4.1.5 二叉树删除

要删除一棵二叉树，需要删除树中所有结点。可以考虑先删除左子树，然后删除右子树，最后删除根结点，这正是后序遍历的过程。所以不难写出如下递归删除函数，代码如下：

```
void Free(TreeNode* t)
{    delete t;   }
void Release(TreeNode *t)
{
    if (!t)
        return;
    Release(t->left);
    Release(t->right);
    Free(t);
}
```

删除树主要需要释放所有结点内存，而与结点释放顺序无关。所以只要能正确访问所有结点，任何方法都可行。下面是非递归删除函数，代码如下：

```
void Release(TreeNode *t)
{
    if(!t)
        return;
    std::stack<TreeNode*> allNodes;
    TreeNode* pNode;
    allNodes.push(t);
    while(!allNodes.empty())
    {
        pNode = allNodes.top();
        allNodes.pop();
        if(pNode->left)
            allNodes.push(pNode->left);
        if(pNode->right)
            allNodes.push(pNode->right);
        delete pNode;
    }
}
```

4.1.6 综合编程实例

1. 二叉搜索树的结构

二叉搜索树或者一棵空树，或者具有下列性质的二叉树： 若它的左子树不空，则左子树上所有结点的值均小于它的根结点的值；若它的右子树不空，则右子树上所有结点的值均大于它的根结点的值；它的左、右子树也分别为二叉搜索树。

给定一系列互不相等的整数，将它们顺次插入一棵初始为空的二叉搜索树，然后对结果树的结构进行描述。你需要判断给定的描述是否正确。例如，将{ 2 4 1 3 0 }插入后，得到一棵二叉搜索树，则陈述句如“2 是树的根”“1 和 4 是兄弟结点”“3 和 0 在同一层上”（指自顶向下的深度相同）、“2 是 4 的双亲结点”“3 是 4 的左孩子”都是正确的；而“4 是 2 的左孩子”“1 和 3 是兄弟结点”都是不正确的。

输入格式：

在第一行输入一个正整数 N（<= 100），随后一行给出 N 个互不相同的整数，整数间以空格分隔，要求将之顺次插入一棵初始为空的二叉搜索树。之后给出一个正整数 M（<= 100），随后 M 行，每行给出一句待判断的陈述句。陈述句有以下六种：

```
"A is the root"，即"A 是树的根"；
"A and B are siblings"，即"A 和 B 是兄弟结点"；
"A is the parent of B"，即"A 是 B 的双亲结点"；
"A is the left child of B"，即"A 是 B 的左孩子"；
"A is the right child of B"，即"A 是 B 的右孩子"；
"A and B are on the same level"，即"A 和 B 在同一层上"。
```

题目保证所有给定的整数都在整型范围内。

输出格式：

对每句陈述，如果正确则输出“Yes”，否则输出“No”，每句占一行。

输入样例 1：

```
5
2 4 1 3 0
8
2 is the root
1 and 4 are siblings
3 and 0 are on the same level
2 is the parent of 4
3 is the left child of 4
1 is the right child of 2
4 and 0 are on the same level
100 is the right child of 3
```

输出样例 2：

```
Yes
Yes
Yes
Yes
Yes
No
No
No
```

这道题来自团体程序设计天梯赛，编号为 L3-016 的二叉搜索树的结构（在线题目网址 https://pintia.cn/problem-sets/994805046380707840/problems/994805047903240192，作者：陈越）。从题目内容来看，这道题难度并不大。但通过率非常低。这道题完全考验程序员在有限时间内构建二叉搜索树及其对应函数的能力。以下是参考代码，代码如下：

```
#include <iostream>
#include <vector>
#include <sstream>

using namespace std;
//定义二叉树结点结构
struct TreeNode
{
    TreeNode(int value=0)
    {
        left = right = NULL;
        v = value;
    }
    TreeNode* left;
    TreeNode* right;
    int v;
};
//定义二叉树类
class BinaryTree
{
public:
    BinaryTree() { m_Root = NULL;}
    ~BinaryTree() { Release();}
    //释放树内存
    void Release()
    {
        if(!m_Root)
            return;
        std::vector<TreeNode*> allNodes;
        TreeNode* pNode;
        allNodes.push_back(m_Root);
        while(!allNodes.empty())
        {
            pNode = allNodes.back();
            allNodes.pop_back();
            if(pNode->left)
                allNodes.push_back(pNode->left);
            if(pNode->right)
                allNodes.push_back(pNode->right);
```

```
            delete pNode;
        }
        m_Root = NULL;
    }
    //是否为根
    bool IsRoot(int value)
    {
        if(m_Root && m_Root->v==value)
            return true;
        else
            return false;
    }
    //搜索结点
    bool SearchNode(int v, TreeNode* &parentNode, TreeNode* &node, int*pDepth=NULL)
    {
        parentNode = NULL;
        node = NULL;
        if(pDepth)
            *pDepth = 0;
        if(m_Root->v==v)
        {
            parentNode = NULL;
            node = m_Root;
            if(pDepth)
                *pDepth = 0;
            return true;
        }
        std::vector<TreeNode*> parents;
        std::vector<int> depths;
        TreeNode* pp;
        int depth;
        parents.push_back(m_Root);
        depths.push_back(0);
        while(!parents.empty())
        {
            pp = parents.back();
            depth = depths.back();
            parents.pop_back();
            depths.pop_back();
            if(pp->left && pp->left->v==v)
            {
                parentNode = pp;
                node = pp->left;
                if(pDepth)
```

```
                    *pDepth = depth+1;
                return true;
            }
            else if(pp->right && pp->right->v==v)
            {
                parentNode = pp;
                node = pp->right;
                if(pDepth)
                    *pDepth = depth+1;
                return true;
            }
            if(pp->left)
            {
                parents.push_back(pp->left);
                depths.push_back(depth+1);
            }
            if(pp->right)
            {
                parents.push_back(pp->right);
                depths.push_back(depth+1);
            }
        }
        return false;
}
//a 和 b 是否为兄弟结点
bool IsSibling(int a, int b)
{
        TreeNode *parentNode, *node;
        if(!SearchNode(a, parentNode, node))
            return false;
        if(!parentNode)
            return false;
        if(parentNode->left && parentNode->right)
        {
            if(parentNode->left->v= =a && parentNode->right->v= =b||
               parentNode->left->v= =b && parentNode->right->v= =a)
               return true;
        }
        return false;
}
//是否是父亲结点
bool IsParent(int parentV, int childV)
{
        TreeNode *parentNode, *node;
```

```
    if(!SearchNode(childV, parentNode, node))
        return false;
    return parentNode && parentNode->v==parentV;
}
//是否是左儿子
bool IsLeftChild(int parentV, int childV)
{
    TreeNode *parentNode, *node;
    if(!SearchNode(parentV, parentNode, node))
        return false;
    return node->left && node->left->v==childV;
}
//是否是右儿子
bool IsRightChild(int parentV, int childV)
{
    TreeNode *parentNode, *node;
    if(!SearchNode(parentV, parentNode, node))
        return false;
    return node->right && node->right->v==childV;
}
//是否在同一层
bool IsSameLevel(int a, int b)
{
    TreeNode *parentNode, *node;
    int depth1, depth2;
    if(!SearchNode(a, parentNode, node, &depth1))
        return false;
    if(!SearchNode(b, parentNode, node, &depth2))
        return false;
    return depth1==depth2;
}
//插入一个值
void InsertValue(TreeNode* &pParent, int v)
{
    if(!pParent)
        pParent = new TreeNode(v);
    else
    {
        if(v<pParent->v)
            InsertValue(pParent->left, v);
        else
            InsertValue(pParent->right, v);
    }
}
```

```
    //根据初始值构建一棵树
    void BuildTree(const std::vector<int>& values)
    {
        Release();
        if(values.empty())
            return;
        size_t i;
        m_Root = new TreeNode(values[0]);
        for(i=1; i<values.size(); ++i)
            InsertValue(m_Root, values[i]);
    }
private:
    TreeNode* m_Root;
};

int main()
{
    int i, n, v1, v2, v;
    std::vector<int> values;
    std::string str, str2;
    std::istringstream iss;
    std::string::size_type pos;

    cin>>n;
    values.resize(n);
    for(i=0; i<n; ++i)
        cin>>values[i];
    //创建搜索二叉树实例
    BinaryTree tree;
    //建树
    tree.BuildTree(values);
    cin>>n;
    //忽略回车
    cin.ignore();
    for(i=0; i<n; ++i)
    {
        //得到一行输入
        std::getline(cin, str);
        iss.clear();
        //将输入打包到字符串输入流
        iss.str(str);
        if(str.find("root")!=std::string::npos) //是否在语句中找到 root 关键字
        {
```

```
        iss>>v;
        if(tree.IsRoot(v))
            cout<<"Yes"<<endl;
        else
            cout<<"No"<<endl;
    }
    else if(str.find("siblings")!=std::string::npos) //是否在语句中找到 sibling 关键字
    {
        iss>>v1>>str2>>v2;
        if(tree.IsSibling(v1, v2))
            cout<<"Yes"<<endl;
        else
            cout<<"No"<<endl;
    }
    else if(str.find("parent of")!=std::string::npos) //是否在语句中找到 parent of 关键字
    {
        iss>>v1;
        pos = str.find_last_of(' ');
        iss.str(std::string(str.begin()+pos, str.end()));
        iss>>v2;
        if(tree.IsParent(v1, v2))
            cout<<"Yes"<<endl;
        else
            cout<<"No"<<endl;
    }
    else if(str.find("left child")!=std::string::npos) //是否在语句中找到 left child 关键字
    {
        iss>>v1;
        pos = str.find_last_of(' ');
        iss.str(std::string(str.begin()+pos, str.end()));
        iss>>v2;
        if(tree.IsLeftChild(v2, v1))
            cout<<"Yes"<<endl;
        else
            cout<<"No"<<endl;
    }
    else if(str.find("right child")!=std::string::npos) //是否在语句中找到 right child 关键字
    {
        iss>>v1;
        pos = str.find_last_of(' ');
        iss.str(std::string(str.begin()+pos, str.end()));
        iss>>v2;
        if(tree.IsRightChild(v2, v1))
            cout<<"Yes"<<endl;
```

```
            else
                cout<<"No"<<endl;
        }
        else if(str.find("same level")!=std::string::npos) //是否在语句中找到 same level 关键字
        {
            iss>>v1>>str2>>v2;
            if(tree.IsSameLevel(v1, v2))
                cout<<"Yes"<<endl;
            else
                cout<<"No"<<endl;
        }
    }
    return 0;
}
```

2. 输出后序遍历

输入一棵二叉树的前序遍历和中序遍历，输出它的后序遍历。

例如，输入：

```
7 3 2 1 5 4 6 9 8 10
1 2 3 4 5 6 7 8 9 10
```


输出：

```
1 2 4 6 5 3 8 10 9 7
```

这是经典的二叉树重构问题。前序遍历的第一个值就是根结点值，所以在中序遍历中找到它，就可以知道左、右子树的前序和中序遍历，代码如下：

```
#include <iostream>   //std::cin, std::cout
#include <algorithm>   //std::find
#include <sstream>     //std::istringstream
#include <cassert>     //assert

using namespace std;
#define N 100

void BuildPost(int * preOrderStr, int* inOrderStr, int n, int* postOrderStr)
{
    if(n<=0)      return;
    //pos 既是左、右子树的分割位置，也是左子树结点个数
    int pos = std::find(inOrderStr, inOrderStr+n, preOrderStr[0]) - inOrderStr;
    //递归构造左子树的后序遍历
    BuildPost(preOrderStr+1, inOrderStr, pos, postOrderStr);
    //递归构造右子树的后序遍历
    BuildPost(preOrderStr+pos+1, inOrderStr+pos+1, n-pos-1, postOrderStr+pos);
    //把根结点添加到后序遍历数组的最后位置
```

```
        postOrderStr[n-1] = preOrderStr[0];
}
int main()
{
        int n1, n2, i;
        int preOrderStr[N], inOrderStr[N], postOrderStr[N];
        std::string str;
        istringstream iss;

        //得到一行输入
        std::getline(std::cin, str);
        //打包这行字符串到字符串流
        iss.str(str);
        n1 = 0;
        while(iss>>i)
                preOrderStr[n1++]=i;

        //得到一行输入
        std::getline(std::cin, str);
        //先清除标志位
        iss.clear();
        //打包这行字符串到字符串流
        iss.str(str);
        n2 = 0;
        while(iss>>i)
                inOrderStr[n2++]=i;
        assert(n1==n2);

        BuildPost(preOrderStr, inOrderStr, n1, postOrderStr);
        for(i=0; i<n1; ++i)
        {
                if(i!=0)
                    cout<<' ';
                cout<<postOrderStr[i];
        }
        cout<<endl;
        return 0;
}
```

需要注意的是，在编写 BuildPost 递归函数时，注意递归构造右子树上的各种索引偏移量。所以读者需要提前确定左、右子树组合，然后对照算法对各项偏移量进行验证。

3. 输出前序遍历

输入一棵二叉树的后序遍历和中序遍历，请输出它的前序遍历。

例如，输入：

```
1 2 4 6 5 3 8 10 9 7
1 2 3 4 5 6 7 8 9 10
```

输出：

```
7 3 2 1 5 4 6 9 8 10
```

这依旧是经典的二叉树重构问题。后序遍历的最后一个值就是根结点值，所以在中序遍历中找到它后，就可以知道左、右子树的后序和中序遍历，代码如下：

```
#include <iostream>   //std::cin, std::cout
#include <algorithm>   //std::find
#include <sstream>    //std::istringstream
#include <cassert>    //assert

using namespace std;
#define N 100
void BuildPre(int * postOrderStr, int* inOrderStr, int n, int* preOrderStr)
{
    if(n<=0)
        return;
    //在中序遍历中找到后序遍历的最后一个值也就是根结点值， pos 既是左、右子树的分割位置，
也是左子树结点个数
    int pos = std::find(inOrderStr, inOrderStr+n, postOrderStr[n-1]) - inOrderStr;
    //后序遍历的最后一个值就是根结点值，也就是前序遍历第一个值
    preOrderStr[0] = postOrderStr[n-1];
    //遍历左子树
    BuildPre(postOrderStr, inOrderStr, pos, preOrderStr+1);
    //遍历右子树
    BuildPre(postOrderStr+pos, inOrderStr+pos+1, n-pos-1, preOrderStr+pos+1);
}
int main()
{
    int n1, n2, i;
    int preOrderStr[N], inOrderStr[N], postOrderStr[N];
    std::string str;
    istringstream iss;

    //得到一行输入
    std::getline(std::cin, str);
    //打包这行字符串到字符串流
    iss.str(str);
    n1 = 0;
    while(iss>>i)
        postOrderStr[n1++]=i;

    //得到另一行输入
```

```
    std::getline(std::cin, str);
    //先清除标志位
    iss.clear();
    //打包这行字符串到字符串流
    iss.str(str);
    n2 = 0;
    while(iss>>i)
        inOrderStr[n2++]=i;
    //断言 n1==n2
    assert(n1==n2);

    BuildPre(postOrderStr, inOrderStr, n1, preOrderStr);
    for(i=0; i<n1; ++i)
    {
        if(i!=0)
            cout<<' ';
        cout<<preOrderStr[i];
    }
    cout<<endl;
    return 0;
}
```


同样需要注意的是，在编写 BuildPre 递归函数时，注意递归构造右子树上的各种索引偏移量。读者依旧需要提前确定左、右子树组合，然后对照算法对各项偏移量进行验证。

4. 构建二叉树

以上两个方法都采用递归方法直接生成二叉树的后序和前序遍历。那么，能否根据前序遍历和中序遍历或者后序遍历和中序遍历真正重构一棵二叉树呢？答案是可以的。只要定义树结点的结构，然后修改已有的递归重构函数即可。下面以后序遍历和中序遍历为例，构造一棵二叉树，代码如下：

```
#include <iostream>    //std::cin, std::cout
#include <algorithm>   //std::find
#include <sstream>     //std::istringstream
#include <cassert>     //assert

using namespace std;
#define N 100

struct TreeNode
{
    TreeNode()
    {
        left = right = NULL;
```

```
            value = 0;
        }
        int value;
        TreeNode *left;
        TreeNode *right;
    };
    void BuildTree(int * postOrderStr, int* inOrderStr, int n, TreeNode *&root)
    {
        if(n<=0)
            return;

        //创建当前子树的根结点
        root = new TreeNode();
        root->value = postOrderStr[n-1];

        //在中序遍历中找到后序遍历的最后一个值也就是根结点值， pos 既是左、右子树的分割位置，
也是左子树结点个数
        int pos = std::find(inOrderStr, inOrderStr+n, postOrderStr[n-1]) - inOrderStr;
        //后序遍历的最后一个值就是根结点值，也就是前序遍历的第一个值

        //遍历构建左子树，所以要传入当前子树的左儿子结点
        BuildTree(postOrderStr, inOrderStr, pos, root->left);
        //遍历构建右子树，所以要传入当前子树的右儿子结点
        BuildTree(postOrderStr+pos, inOrderStr+pos+1, n-pos-1, root->right);
    }

    //打印前序遍历结果，变量 bFirst 用来控制是否输出空格
    void PreOrder(bool bFirst, const TreeNode* root)
    {
        if(!root)
            return;
        if(!bFirst)
            cout<<" ";
        cout<<root->value;
        PreOrder(false, root->left);
        PreOrder(false, root->right);
    }
    //采用后序遍历方法释放内存
    void FreeTree(TreeNode* root)
    {
        if(!root)
            return;
        FreeTree(root->left);
        FreeTree(root->right);
```

```
        delete root;
    }

    int main()
    {
        int n1, n2, i;
        int inOrderStr[N], postOrderStr[N];
        std::string str;
        istringstream iss;

        //得到一行输入
        std::getline(std::cin, str);
        //打包这行字符串到字符串流
        iss.str(str);
        n1 = 0;
        while(iss>>i)
            postOrderStr[n1++]=i;

        //得到另一行输入
        std::getline(std::cin, str);
        //先清除标志位
        iss.clear();
        //打包这行字符串到字符串流
        iss.str(str);
        n2 = 0;
        while(iss>>i)
            inOrderStr[n2++]=i;
        //断言 n1==n2
        assert(n1==n2);

        TreeNode *root=NULL;
        BuildTree(postOrderStr, inOrderStr, n1, root);
        PreOrder(true, root);
        cout<<endl;
        FreeTree(root);
        return 0;
    }
```

程序中首先定义 TreeNode 结构，然后依旧通过递归方法构建树。递归函数需要传入结点指针的引用，因为内存是在递归程序的内部分配的，所以必须传入指针的引用或者指针的指针。最后不要忘记用后序遍历法释放内存。

4.2 图

图是数据结构和算法中一个重要内容。图通常表示成 ***G***= (***V***, ***E***)的集合形式，其中 ***V*** 是顶点(vertex)集合，***E*** 是边(edge)集合。***E*** 中的每一条边连接 ***V*** 中两个不同的顶点，所以边可以用 $e_k = \langle v_i, v_j \rangle$ 形式来表示，其中 v_i 和 v_j 是 ***V*** 中两个顶点。

图通常被分为三种类型：无向图、有向图和加权图。图的常用表示方法有邻接矩阵、邻接链表和邻接压缩表方法。图的标准搜索方法有宽度优先搜索和深度优先搜索两种。

4.2.1 图的基本操作

涉及图的数据结构通常包含如下几个基本操作。

- adjacent(G, x, y): 图 ***G*** 中是否有一条边从顶点 *x* 指向顶点 *y*。
- neighbors(G, x): 列出所有从顶点 *x* 出发的边的另外一个顶点。
- add_vertex(G, x): 如果顶点 *x* 不存在，则将 *x* 添加到图中。
- remove_vertex(G, x): 删除顶点 *x*。
- add_edge(G, x, y): 如果边<*x*,*y*>不存在，则添加这条新边。
- remove_edge(G, x, y): 如果边<*x*,*y*>存在，则删除。
- get_vertex_value(G, x): 得到和顶点 *x* 关联的一个数值。
- set_vertex_value(G, x, v): 将数值 *v* 关联到顶点 *x*。
- get_edge_value(G, x, y): 得到和边<*x*, *y*>关联的一个数值。
- set_edge_value(G, x, y, v): 将数值 *v* 关联到边<*x*, *y*>。

对于在线测试程序和部分简单算法而言，可能并不需要手动写出如上的各种操作函数。但在实际涉及图数据的软件开发中，上述操作必不可少。

4.2.2 图的表示方法

图的常用表示方法有三种：邻接矩阵、邻接链表和邻接压缩表。

1. 邻接矩阵（Adjacency Matrix）

邻接矩阵是一个二维矩阵，其中行表示边的起始顶点，列表示边的终止顶点。例如，对于有 *n* 个顶点的图 ***G***，它的邻接矩阵就是一个 $n \times n$ 矩阵，矩阵中每个元素值为 0 或 1。对于无向图来说，这个矩阵一定是对称的。邻接矩阵表示方法，简单明了，因此是图的简单算法中的常用表示形式。

2. 邻接链表（Adjacency List）

邻接链表中的顶点存储在一个线性结构中。它的每个顶点结构都记录了自身所有的邻接顶点或邻接边。在实际工程中，往往还需要将所有边也存储在另外一个线性结构中，每条边都记录其关联的两个顶点，甚至还记录和该边相连的所有邻接边。

3. 邻接压缩表（Adjacency Packed List）

邻接压缩表需要两个一维数组，其中一个数组 *A* 用来存放所有的邻接顶点，另一个数组 *B* 存放每个顶点的邻接顶点数目。首先，将所有邻接于顶点 1 的顶点加入数组 *A*，然后将所有邻接于顶点 2 的顶点加入数组 *A*，这样一直进行下去。数组 *B* 中每项数据存储每个顶点的邻接顶点数目。

4.2.3 综合编程实例

1. 打印连通集合

给定一个无向图 **G=（V, E）**，其中顶点有 *v* 个（编号从 0 到 *v*-1)，边有 *e* 条，请用宽度优先搜索和深度优先搜索打印该无向图的连通集合，遍历时要求从编号小的顶点开始。第一行输入 *v* 和 *e*，接着输入 *e* 条边的两个顶点。输出的每个连通集合占据一行，连通集合的顶点之间用一个空格分隔，行尾不要有多余空格。

例如，输入：

```
9 7
1 2
8 2
0 3
3 6
5 3
3 7
7 5
```

输出：

```
0 3 5 7 6
1 2 8
4
0 3 5 6 7
1 2 8
4
```

这是经典的 DFS 和 BFS 问题。BFS 采用队列结构访问顶点，而 DFS 则采用堆栈结构。示例代码如下：

```
#include <iostream>
#include <queue> //std::queue

#define MAX_N 50

//采用邻接矩阵表示图。因为是全局变量，所以默认值都是 0，表示初始图中没有任何边
bool graph[MAX_N][MAX_N];

//宽度优先遍历
```

```
void BFS(int v)
{
    int i, j, seed;
    //数组表示顶点是否已经被访问，访问的顶点不再进队列
    bool bVerticesTakens[MAX_N]={false};
    //当前行的第一个被打印的数
    bool bFirst;
    for(i=0; i<v; ++i)
    {
        if(!bVerticesTakens[i])
        {
            //设置访问的标志位
            bVerticesTakens[i] = true;
            //宽度优先搜索采用队列结构
            std::queue<int> seeds;
            //前行的第一个被打印的数
            bFirst = true;
            //先把当前顶点加入到队列中
            seeds.push(i);
            //队列不为空
            while(!seeds.empty())
            {
                //从队列前端取出种子顶点
                seed = seeds.front();
                seeds.pop();
                if(!bFirst)
                    std::cout<<" ";
                else
                    bFirst=false;
                std::cout<<seed;
                //寻找和种子点邻接并且尚未被访问的顶点
                for(j=0; j<v; ++j)
                {
                    //和顶点 seed 邻接且尚未被访问的顶点
                    if(graph[seed][j] && !bVerticesTakens[j])
                    {
                        //设置访问标志位
                        bVerticesTakens[j] = true;
                        //加入到队列中
                        seeds.push(j);
                    }
                }
            }
```

```
                std::cout<<"\n";
            }
        }
    }

    //深度优先遍历的递归函数，传入的是当前连通集合的第一个种子顶点 seed
    //而 bTakens 是顶点访问标志位数组， bFirst 表示是否是当前连通集合的第一个元素
    void DFS_Interator(int seed, int v, bool* bTakens, bool bFirst)
    {
        int i;
        if(!bFirst)
            std::cout<<" ";
        std::cout<<seed;
        for(i=0; i<v; ++i)
        {
            //和顶点 seed 邻接且尚未被访问的顶点
            if(graph[seed][i] && !bTakens[i])
            {
                bTakens[i] = true;
                //深度递归遍历
                DFS_Interator(i, v, bTakens, false);
            }
        }
    }

    //深度优先遍历
    void DFS(int v)
    {
        bool bTakens[MAX_N]={false};
        int i;
        //遍历尚未被访问的连通集
        for(i=0; i<v; ++i)
        {
            //找到该集合的第一个种子
            if(!bTakens[i])
            {
                //设置访问标志位
                bTakens[i] = true;
                //深度优先访问该连通集合
                DFS_Interator(i, v, bTakens, true);
                std::cout<<"\n";
            }
        }
```

```
}

int main()
{
    int i, v, e, v1, v2;
    std::cin>>v>>e;
    for(i=0; i<e; ++i)
    {
        std::cin>>v1>>v2;
        graph[v1][v2]=true;
        graph[v2][v1]=true;
    }
    DFS(v);
    BFS(v);
}
```

2. 计算八连通区域块数量

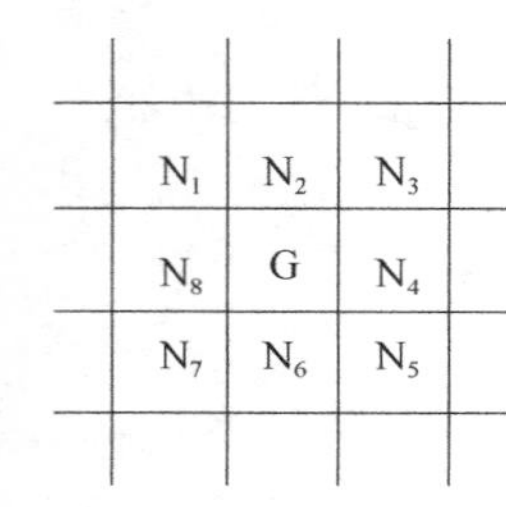

图 4-5　八连通示意图

八连通区域是指从二维平面区域内每个像素出发，可通过八个方向，即上、下、左、右、左上、右上、左下、右下这八个方向的移动的组合，在不超出区域的前提下，到达区域内的任意像素。八连通示意图如图 4-5 所示。

如图 4-5 所示，像素点 G 除与 N_2，N_4，N_6，N_8 像素连通外，还和像素点 N_1，N_3，N_5，N_7 也是连通的。下面就要计算一幅黑白图像中的黑色八连通区域块的数量。假设图像是 $n \times n$ 个像素点，输入第一行是 n，接着输入 n 行数据，其中 1 表示黑色，0 表示白色。

例如，输入：

```
8
1 0 1 0 1 1 0 1
0 1 0 0 1 0 0 0
1 0 1 0 1 1 0 1
0 0 0 0 0 0 0 0
1 0 0 0 0 0 0 0
1 1 1 0 1 0 0 0
1 0 0 0 1 1 0 0
1 0 0 0 0 0 1 1
```

对应的图像如图 4-6 所示。

输出：

```
Region cout = 6
```

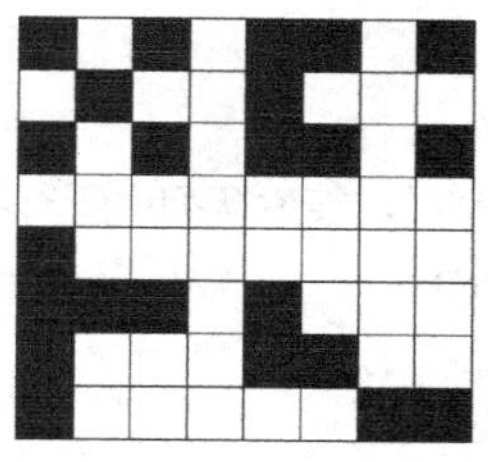

图 4-6 对应的图像

以上示例代码如下：

```
#include <iostream>
#include <queue>
using namespace std;

#define N 100

//为了节省空间，这里用 bool 表示黑白格子，由于是全局变量，所以初始都为 0，表示白色
bool data[N][N];
//访问标志位，避免种子重复进入队列
bool bVisits[N][N];
int main()
{
    int i, j, n, k, x, y, regionCount=0; //八连块的数目
    std::pair<int, int> seed; //种子，这里用 pair 表示输入数据部分的行号和列号
    /* i 从 1 开始，是因为在图的四周加了一层栅格（虚拟的白格子），从而避免出界判断 */
    int xSteps[8]={-1, 0, 1, 1, 1, 0, -1, -1};
    int ySteps[8]={-1, -1, -1, 0, 1, 1, 1, 0};
    cin>>n;
    for(i=1; i<=n; ++i)
    {
        for(j=1; j<=n; ++j)
            cin>>data[i][j];
    }
    /* i 从 1 开始，是因为在图的四周加了一层栅格（虚拟的白格子），从而避免出界判断 */
    for(i=1; i<=n; ++i)
    {
        for(j=1; j<=n; ++j)
        {
            //找到一个种子(没有被访问的黑格子）
            if(!bVisits[i][j] && data[i][j])
            {
                ++regionCount; //八连块的数目
                std::queue<std::pair<int, int> > seeds; //种子队列
                seeds.push(std::make_pair(i, j));
```

```
                    bVisits[i][j] = true;

                    //当种子队列不为空时，搜索队列中所有种子的相邻黑格子
                    while(!seeds.empty())
                    {
                        seed = seeds.front();
                        seeds.pop();
                        //8 个相邻的方向
                        for(k=0; k<8; ++k)
                        {
                            x = seed.first + xSteps[k];
                            y = seed.second + ySteps[k];
                            //找到相邻的另一个种子(没有被访问的黑格子）
                            if(!bVisits[x][y] && data[x][y])
                            {
                                seeds.push(std::make_pair(x, y));
                                bVisits[x][y] = true;
                            }
                        }
                    }
                }
            }
        }
        cout<<"Region cout = "<<regionCount<<endl;
        return 0;
}
```

上述代码采用了宽度优先搜索方法。为了避免出界判断，这里在实际图像的四周包围了一圈白色虚拟栅栏。这意味对于 8×8 的图像，需要 10×10 的数组存储图像。另外，对于种子队列，这里使用 std::pair 来表示种子的行号和列号，当然也可以将行号和列号编码成整数 d = row*N+col。

3. 迷宫问题

在 $M \times N$ 单元格组成的迷宫里，需要找一条从起点到终点的最短路径。$M \times N$ 迷宫用一个二维数组表示，其中 1 表示墙，0 表示可以走的路，只能横着走或竖着走，不能斜着走。

这是广度优先搜索算法的一道经典例题。算法要点是要多分配一个 $M \times N$ 的方向数组，用来记录每个格子是从哪个方向走过来的。当 BFS 遍历整个迷宫后，该方向数组会被填满方向信息。此时只需要从该方向数组的终点反向往回推到起点，从而得到一系列格子坐标，最后输出这些坐标即可。迷宫路径方向信息如图 4-7 所示。

图 4-7　迷宫路径方向信息

迷宫问题示例代码如下：

```
#include <iostream>
#include <queue>
using namespace std ;

#define MAX_N 104
bool maze[MAX_N][MAX_N], bVisits[MAX_N][MAX_N];
int lastDirs[MAX_N][MAX_N]; //任意位置的移动方向
std::pair<int, int> fatherPos[MAX_N][MAX_N]; //父结点编号
//(x,y)是起点坐标
void BFS(int x, int y, int n, int m)
{
    std::pair<int, int> seed;
    //种子队列
    std::deque<std::pair<int, int> > seeds;
   //利用方向移动数组使得代码很简捷
    int dx[4] = {-1,1,0,0} ; //up, down, left, right
    int dy[4] = {0,0,-1,1} ;
    int nx, ny, i;

    bVisits[x][y] = true;
    seed = std::make_pair(x, y);
   //起点的父亲位置就设置成自己，后续打印路径时会用到这个特征
   fatherPos[x][y] = seed;
    seeds.push_back(seed) ;
    while (!seeds.empty())
    {
        seed = seeds.front() ;
        seeds.pop_front() ;
        x = seed.first;
```

```
            y = seed.second;
        for (i = 0 ; i < 4 ; ++ i)
        {
            nx = x + dx[i] ;
            ny = y + dy[i] ;
            //利用虚拟的栅栏边界方法，避免了越界检查
            //是通路而且没有被访问
            if (maze[nx][ny] && !bVisits[nx][ny])
            {
                bVisits[nx][ny] = true ; //设置访问标志位
                fatherPos[nx][ny] = seed ; //记录来路
                //添加新种子
                seeds.push_back(std::make_pair(nx, ny)) ;
                lastDirs[nx][ny] = i ; //记录来路的方向
            }
        }
    }
}
//打印最短路径的行走方向，(x,y)是终点坐标
void PrintPath(int x, int y)
{
    //dirs 存储一条从终点到起点的反向行走路径
    std::vector<int> dirs;
    char name[] = "UDLR" ;
    while (true)
    {
        int fx = fatherPos[x][y].first;
        int fy = fatherPos[x][y].second;
        if (fx == x && fy == y) //碰到入口
            break ;
        dirs.push_back(lastDirs[x][y]);
        x = fx ;
        y = fy ;
    }
    //打印行走方向
    for(int i=(int)dirs.size()-1; i>=0; --i)
        cout << name[dirs[i]] ;
}

int main()
{
    int i, j, n, m, startX, startY, endX, endY ;
    cin >> n >> m >> startX >> startY >> endX >> endY ;
```

```
    //从 1 开始，利用虚拟删格技术，避免越界检查
    for (i = 1 ; i <= n ; ++ i)
    {
        for (j = 1 ; j <= m ; ++ j)
            cin >> maze[i][j] ;
    }
    BFS(startX+1,startY+1, n, m);
    PrintPath(endX+1,endY+1);
    return 0 ;
}
```

第 5 章　递归与分治

递归（Recursion）是一个函数在其定义中直接或间接调用自身的一种方法。递归方法通常把一个复杂的问题转化为一个与原问题相似但规模较小的问题来求解，从而大大减少了程序的代码量。递归程序一般需要有边界条件、前进段和返回段三部分。为了防止递归死循环，一般需要将边界条件的判断代码放在函数开始部分。

分治（Divide and Conquer）是分而治之的简称。它是把一个复杂的问题分成多个子问题，再把子问题分成更小的子问题，直到最后子问题可以直接求解，最后需要将子问题的解结果进行合并。所以分治的求解过程：任务分解、子问题求解及最后结果合并。

5.1　汉诺塔

有三根柱子 A、B 和 C。A 柱上有 N 个（$N>1$）穿孔圆盘，盘的尺寸由下到上依次变小。要求按下列规则将所有圆盘移至 B 柱：每次只能移动一个圆盘并且大盘不能放在小盘上面。问：如何将 A 柱上的所有圆盘都移动到 C 柱上？最少要移动多少次？

设移动盘子数为 n，为了将这 n 个盘子从 A 柱移动到 C 柱，可以做以下三步工作：

（1）以 C 柱为中介，从 A 柱上将 1 至 n～1 号盘移至 B 柱。

（2）将 A 柱上剩下的最后第 n 号盘移至 C 柱。

（3）以 A 柱为中介，将 B 柱上的 1 至 n～1 号盘移至 C 柱。

根据以上规则，不难写出如下递归函数，代码如下：

```
#include <iostream>
using namespace std;

void Hanoi(int n, char src, char media, char des, int &moveCount)
{
    if(n==1)
    {
        cout<<src<<"->"<<des<<endl;
        ++moveCount;
        return ;
    }
```

```
        Hanoi(n-1, src, des, media, moveCount);
        cout<<src<<"->"<<des<<endl;
        ++moveCount;
        Hanoi(n-1, media, src, des, moveCount);
}

int main()
{
        int moveCount, diskNum=3;
        while(true)
        {
                cout<<"Please input disk number: ";
                cin>>diskNum;
                if(diskNum<1)
                        break;
                moveCount = 0;
                Hanoi(diskNum, 'A', 'B', 'C', moveCount);
                cout<<"Steps count = "<<moveCount<<endl<<endl;
        }
        return 0;
}
```


对于递归函数，通常建议把递归退出判断条件的 if 语句写在递归函数的第一行。

5.2 子串组合

从长度为 n 的字符串中选择 m 个字符组成子串，求子串的所有结果。

这是一个典型的组合问题，子串个数显然是 C_n^m 。对于字符串中第一个字符，有两种选择。

（1）把这个首字符放到当前组合中，接下来需要在剩下的 n~1 个字符中选择 m~1 个字符。

（2）不把这个首字符放到当前组合中，接下来需要在剩下的 n~1 个字符中选择 m 个字符。

根据以上规则，可以写出如下递归函数，代码如下：

```
#include <cstring>
#include <cstdio>
#include <cstdlib>

void Combination(const char *str, int m, char *subStr, int subStrLen)
{
        //不需要再选择字符，已经得到子串
        if(m == 0)
        {
                int i;
                for(i=0; i<subStrLen; ++i)
```

```
                printf("%c", subStr[i]);
            printf("\n");
            return;
        }

        //还没有选够子串，但已经遍历到字符串尾部，所以要退回
        if(*str == '\0')
            return;

        //情况 1：将当前首字符加到子串尾部，所以需要从剩下的字符中选择 m-1 个
        subStr[subStrLen++]=*str;
        //递归调用
        Combination(str + 1, m - 1, subStr, subStrLen);
        //情况 2：不将当前首字符加到子串尾部，所以需要在剩下的字符中选择 m 个
        subStrLen--;
        //递归调用
        Combination(str + 1, m, subStr, subStrLen);
    }

    void CombinationAll(const char* str)
    {
        if(str == NULL)
            return;

        int n = strlen(str);
        //为子串分配足够的空间
        char *subStr = (char*)malloc(sizeof(char)*(n+1));
        int subStrLen;

        for(int m = 1; m <= n; ++ m)
        {
            //设置一开始子串长度为 0
            subStrLen = 0;
            //从长度为 n 的字符串中选择 m 个字符构成子串
            Combination(str, m, subStr, subStrLen);
        }
        free(subStr);
    }

    int main()
    {
        char str[64];
        fgets(str, 64, stdin);
```

```
    //set \n to \0
    int length = strlen(str);
    str[length-1] = '\0';

    CombinationAll(str);
    return 0;
}
```

CombinationAll 是将子串长度从 1 到 n 的所有组合都打印。

5.3 数组组合

某软件公司有 n 个不同名字的程序员，现需要从中选择 m 个参加即将开始的项目，请问有多少种组合情况？打印名字时请按字典顺序打印。

例如，输入：

```
Jim Tom Allen Beck Kitty Charlie
3
```

输出：

```
Allen Beck Charlie
Allen Beck Jim
Allen Beck Kitty
Allen Beck Tom
Allen Charlie Jim
Allen Charlie Kitty
Allen Charlie Tom
Allen Jim Kitty
Allen Jim Tom
Allen Kitty Tom
Beck Charlie Jim
Beck Charlie Kitty
Beck Charlie Tom
Beck Jim Kitty
Beck Jim Tom
Beck Kitty Tom
Charlie Jim Kitty
Charlie Jim Tom
Charlie Kitty Tom
Jim Kitty Tom
```

以上 n=6，m =3 所以组合数显然是 C_6^3=20 个。此题的解题思路和第 5.2 节子串组合问题一样。对于数组中的第一个元素，有两种选择。

（1）把第一个元素放到当前组合中，接下来需要在剩下的 $n-1$ 个元素中选择 $m-1$ 个元素。

（2）不把第一个元素放到当前组合中，接下来需要在剩下的 $n-1$ 个元素中选择 m 个元素。

另外，程序要求按字典顺序打印组合中的名字。这里有两个解决方法：

（1）先组合，然后对组合内的元素进行字典排序。

（2）先对数组进行字典排序，然后组合。

显然应该选择第二种方案，通过数组预先排序的方法，从而避免后续大量的排序操作。

基于上述描述，不难写出如下递归函数，代码如下：

```
#include <iostream> //for std::cin, std::cout
#include <string>      //for std::string
#include <vector>      //for std::vector
#include <sstream>    //for std::istringstream
#include <cassert>    //for assert
#include <algorithm> //for std::sort

using namespace std;
//所有名字存储在全局 vector 对象中
std::vector<std::string> gNames;

//用 curID==n 表示数组越界
void Combination(int curID, int n, int m, int *groupIDs, int groupIDNum)
{
    //已经得到子数组
    if(m == 0)
    {
        int i;
        for(i=0; i<groupIDNum; ++i)
        {
            if(i!=0)
                cout<<" ";
            cout<<gNames[groupIDs[i]];
        }
        cout<<endl;
        return;
    }

    //还没有选够元素，但是已经到数组的尾部，所以要退回
    if(curID == n)
        return;

    //情况 1：将当前数组的第一个元素加到子数组尾部，所以需要从剩下的元素中选择 m－1 个元素
    groupIDs[groupIDNum++]=curID;
    //递归调用
```

```
        Combination(curID + 1, n, m - 1, groupIDs, groupIDNum);
        //情况 2：不将当前数组的第一个元素加到子数组尾部，所以需要在剩下的字符中选择 m 个
        groupIDNum--;
        //递归调用
        Combination(curID + 1, n, m, groupIDs, groupIDNum);
    }

    void CombinationAll()
    {
        int n = gNames.size();
        if(n==0)
            return;
        int *groupIDs = new int[n];
        int groupIDNum;

        for(int m = 1; m <= n; ++ m)
        {
            //设置一开始子串长度为 0
            groupIDNum = 0;
            //从长度为 n 的字符串中选择 m 个字符构成子串
            Combination(0, n, m, groupIDs, groupIDNum);
        }
        delete[] groupIDs;
    }

    void Combination(int n, int m)
    {
        if(n<=0||m<=0)
            return;
        assert(m<=n);
        int *groupIDs = new int[n];
        int groupIDNum;
        //设置一开始子串长度为 0
        groupIDNum = 0;
        //从长度为 n 的字符串中选择 m 个字符构成子串
        Combination(0, n, m, groupIDs, groupIDNum);
        delete[] groupIDs;
    }

    //Sample: Jim Tom Allen Beck Kitty Charlie
    int main()
    {
        std::string str;
```

```
    std::istringstream iss;
    int m;

    gNames.clear();
    std::getline(std::cin, str);
    iss.str(str);
    while(iss>>str)
    {
        gNames.push_back(str);
    }
    std::cin>>m;

    //先对名字数组按字典顺序排序
    std::sort(gNames.begin(), gNames.end());
    int n = gNames.size();
    Combination(n, m);
    //所有组合情况
    // CombinationAll();
    return 0;
}
```


相比第 5.2 节子串组合方法，上述组合递归方法采用数组表示，所以更通用。

5.4 格子排列

在 $n \times n$ 的二维白色表格内，使用黑色涂 n 个格子，使得任意两个黑色格子不得处在同一行或同一列。格子排列示意图如图 5-1 所示，表示 3×3 表格的 6 种符合条件的涂色方法。

 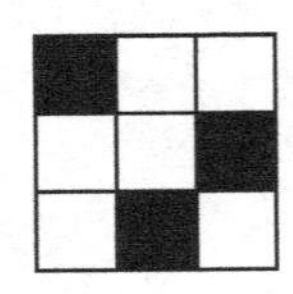 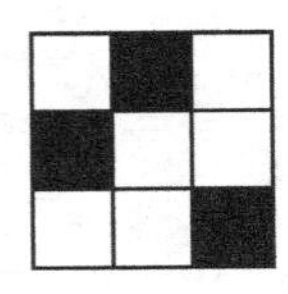   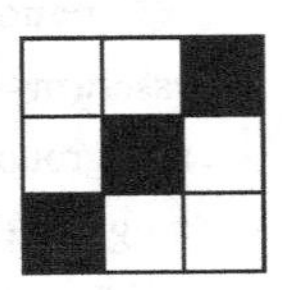

图 5-1 格子排列示意图

显然这是一个排列问题，对于 3×3 表格，解法个数是 $P_3^3=6$。由于任意两个黑色的格子不能处在同一列，那么需要每个黑色的格子都占一列。所以在排列的时候只需记住已排列的每列黑格子的行号，这里使用了一个一维数组 rowIDs 存储每列格子的行号。另外，如何判断每行是否已经被某个黑格子占据，这里用一个一维数组 bRowTakens 表示某行是否已被占据，如果当前行没被占据，则设置占据标志，递归调用下一列黑格子，之后再取消当前行的占据标志。基于上述内容，可以用递归方法解决排列问题，代码如下：

```
#include <iostream> //std::cin, std::cout
#include <cstring> //memset
```

```
using namespace std;

void Permutation(int curColumnID, int n, bool* bRowTakens, int* rowIDs)
{
    int i;
    //当前列号已经达到 n，该排列结束
    if(curColumnID==n)
    {
        for(i=0; i<n; ++i)
        {
            if(i!=0)
                cout<<" ";
            cout<<rowIDs[i];
        }
        cout<<endl;
        return;
    }
    //将当前列的黑格子，试探性放在每一行
    for(i=0; i<n; ++i)
    {
        //如果当前行已被占据
        if(bRowTakens[i])
            continue;
        //当前行没被占据，那么占据它
        bRowTakens[i] = true;
        //设置当前列的行号
        rowIDs[curColumnID] = i;
        //递归调用下一列黑格子
        Permutation(curColumnID+1, n, bRowTakens, rowIDs);
        //取消当前行的占据标志
        bRowTakens[i] = false;
    }
}

void Permutation(int n)
{
    bool *bRowTakens;
    int *rowIDs;
    //当前所有行都没有被占据
    bRowTakens = new bool[n];
    memset(bRowTakens, 0, sizeof(bool)*n);
    //记住每列黑格子所在的行号
    rowIDs = new int[n];
```

```
    //0 表示从第一列黑格子开始排列
    Permutation(0, n, bRowTakens, rowIDs);
    delete[] bRowTakens;
    delete[] rowIDs;
}

int main()
{
    int n;
    cin>>n;
    if(n<=0)
        return 0;
    Permutation(n);
    return 0;
}
```

以上方法是用 bRowTakens 数组检测行是否已经被黑格子占据，是否有更直接的方式检测或者避免检测呢？解决方案：先将 rowIDs 的 n 个数字分别用 0 至 n－1 初始化，接下来要做的事情就是对数组 rowIDs 进行全排列。由于是用不同的数字初始化数组中的数字，因此任意两个黑格子肯定不在同一行。要对这个数组进行全排列，可以考虑先将第一个元素固定，然后对后面的 n－1 个元素进行排列。当后面 n－1 元素排列得出后，可以将第一个元素和第二个元素交换，再固定第一个元素，求后面 n－1 个元素的排列。归纳起来，在全排列的递归中，需要将当前黑格子所在的行号和之后黑格子所在的行号进行两次交换。

解决方案代码如下：

```
void Permutation2(int* rowIDs, int n, int curLen)
{
    int i;
    //当前列号已经达到 n，该排列结束
    if(curLen==n)
    {
        for(i=0; i<n; ++i)
        {
            if(i!=0)
                cout<<" ";
            cout<<rowIDs[i];
        }
        cout<<endl;
        return;
    }
    for(i=curLen; i<n; ++i)
    {
        std::swap(rowIDs[i], rowIDs[curLen]);
        Permutation2(rowIDs, n, curLen+1);
        std::swap(rowIDs[i], rowIDs[curLen]);
```

```
        }
    }

    void Permutation2(int n)
    {
        int *rowIDs, i;
        //记住每列黑格子所在的行号
        rowIDs = new int[n];
        for(i=0; i<n; ++i)
            rowIDs[i] = i;
        //0 表示从第一列黑格子开始排列
        Permutation2(rowIDs, n, 0);
        delete[] rowIDs;
    }
```

还有没有更简单的方法解决全排列问题？有！用 STL 的 next_permutation。std::next_permutation 有两个重载函数：

```
template <class BidirectionalIterator>
  bool next_permutation (BidirectionalIterator first,
                         BidirectionalIterator last);
template <class BidirectionalIterator, class Compare>
  bool next_permutation (BidirectionalIterator first,
                         BidirectionalIterator last, Compare comp);
```

该函数的作用是将当前序列的元素按照升序或自定义排序方法重排至下一个排列。如果有下一个更高的排列，函数就排成更高的排列，并且返回 true。如果没有（已经到了最高排列），则函数将序列排成第一个排列，并且返回 false。代码如下：

```
    void Permutation3(int n)
    {
        int *rowIDs, i;
        //n 个元素
        rowIDs = new int[n];
        for(i=0; i<n; ++i)
            rowIDs[i] = i;
        //rowIDs 已经是排序的，但为了通用性，强调调用 next_permutation 之前，如果要全
    排序，则一定要预先排序序列元素
        std::sort(rowIDs, rowIDs+n);
        gCount = 0;
        //0 表示从第一列黑格子开始排列
        do
        {
            ++gCount;
            for(i=0; i<n; ++i)
            {
```

```
                if(i!=0)
                    cout<<" ";
                cout<<rowIDs[i];
            }
            cout<<endl;
        }
        while(next_permutation(rowIDs, rowIDs+n));
        delete[] rowIDs;
        cout<<"Total permutation num: "<<gCount<<endl;
}
```

5.5 八皇后

在 8×8 的国际象棋上摆放八个皇后，使其不能相互攻击，即任意两个皇后不得处在同一行、同一列或同一对角斜线。求出所有符合上述要求的摆法。

很显然八皇后问题是格子排列问题的特例，多出的一个限制条件是任意两个皇后不能在同一对角线上。对于数组中两个元素下标 i 和 j 而言，i-j==rowIDs[i]-rowIDs[j]或者 j-i==rowIDs[i]-rowIDs[j]就是对角线的判断条件。示例代码如下：


```
#include <iostream> //std::cin, std::cout
#include <cstring> //memset

using namespace std;

int gNumSolutions = 0;

bool SameDiagnoal(int *rowIDs, int n)
{
    int i, j;
    for(i=0; i<n; ++i)
    {
        for(j=i+1; j<n; ++j)
        {
            if(i-j= =rowIDs[i]-rowIDs[j] || i-j= =rowIDs[j]-rowIDs[i])
                return true;
        }
    }
    return false;
}
void Permutation(int* rowIDs, int n, int curLen)
{
    int i;
```

```
        if(curLen==n)
        {
            if(SameDiagnoal(rowIDs, n))
                return;
            ++gNumSolutions;
            for(i=0; i<n; ++i)
            {
                if(i!=0)
                    cout<<" ";
                cout<<rowIDs[i];
            }
            cout<<endl;
            return;
        }
        for(i=curLen; i<n; ++i)
        {
            std::swap(rowIDs[i], rowIDs[curLen]);
            Permutation(rowIDs, n, curLen+1);
            std::swap(rowIDs[i], rowIDs[curLen]);
        }
    }

    void Permutation(int n)
    {
        int *rowIDs, i;
        rowIDs = new int[n];
        for(i=0; i<n; ++i)
            rowIDs[i] = i;
        gNumSolutions = 0;
        Permutation(rowIDs, n, 0);
        delete[] rowIDs;
        cout<<"Num solutions: "<<gNumSolutions<<endl;
    }

    int main()
    {
        int n;
        cin>>n;
        if(n<=0)
            return 0;
        Permutation(n);
        return 0;
    }
```

八皇后问题最终结果是一共有 92 组解。

5.6 循环赛日程安排

有 $n=2^k$ 个选手参加网球循环赛，要求设计一个满足如下要求的比赛日程表。

（1）每个选手必须与其他 $n-1$ 个选手比赛一次。

（2）每个选手一天只能比赛一次。

按上述要求，可以将比赛日程表设计成一个 n 行 $n-1$ 列的二维表，其中第 i 行第 j 列的元素表示第 i 个选手在第 j 天比赛的对手号码。

采用分治策略，可将所有参加比赛的选手分成两部分，$n=2^k$ 个选手的比赛日程表就可以通过 $n=2^{k-1}$ 个选手的比赛日程表来决定。递归执行任务，直到只剩下两个选手，之后将结果进行合并。

表 5-1 展示了 2 人比赛日程表，表 5-2 展示了 4 人比赛日程表，表 5-3 展示了 8 人比赛日程表。

表 5-1　2 人比赛日程表

选手	第一天
1 号	2 号
2 号	1 号

表 5-2　4 人比赛日程表

选手	第一天	第二天	第三天
1 号	2 号	3 号	4 号
2 号	1 号	4 号	3 号
3 号	4 号	1 号	2 号
4 号	3 号	2 号	1 号

表 5-3　8 人比赛日程表

选手	第一天	第二天	第三天	第四天	第五天	第六天	第七天
1 号	2 号	3 号	4 号	5 号	6 号	7 号	8 号
2 号	1 号	4 号	3 号	6 号	5 号	8 号	7 号
3 号	4 号	1 号	2 号	7 号	8 号	5 号	6 号
4 号	3 号	2 号	1 号	8 号	7 号	6 号	5 号
5 号	6 号	7 号	8 号	1 号	2 号	3 号	4 号
6 号	5 号	8 号	7 号	2 号	1 号	4 号	3 号
7 号	8 号	5 号	6 号	3 号	4 号	1 号	2 号
8 号	7 号	6 号	5 号	4 号	3 号	2 号	1 号

在表 5-2 中，共有 4 位选手。其中 1、2 号选手第一天的比赛日程表位于左上角的灰色子表格部分，3、4 号选手第一天的比赛日程表位于左下角的白色子表格部分。而后两天的日程表可以将左上角的子表按其对应位置抄到右下角的子表，左下角的子表可以按其对应位置抄到右上角的子表。

在表 5-3 中，共有 8 位选手。其中左上角的子表（灰色背景）是选手 1 至选手 4 的前三天的比赛日程，左下角是选手 5 至选手 8 前三天的比赛日程。此后四天的比赛日程，就是分别将左上角子表按其对应位置抄到右下角，将左下角的子表按其对应位置抄到右上角。这样就完成了比赛日程的安排。

在每次迭代求解的过程中，可以看作将比赛日程四个部分。

（1）左上角子表：左上角子表是前 2^{k-1} 个选手的前半程的比赛日程。

（2）左下角子表：左下角子表是余下的 2^{k-1} 个选手的前半程的比赛日程。左下角子表与左上角子表的对应关系是左下角子表中对应元素等于左上角子表对应元素加 2^{k-1}。

（3）右上角子表：等于左下角子表的对应元素。

（4）右下角子表：等于左上角子表的对应元素。

以上描述部分，参考代码如下：

```
#include <iostream>
#include <iomanip>

using namespace std;
void Agenda(int k)
{
    int i, j,   subTableLen, ii, jj;
    int n= 1<<k;

    int **tables = new int*[n];
    for(i=0; i<n; ++i)
        tables[i] = new int[n];
    //初始化两人比赛问题的日程表
    tables[0][0] = 1;
    tables[0][1] = 2;
    tables[1][0] = 2;
    tables[1][1] = 1;
    subTableLen = 2;
    for(i=2; i<=k; ++i)
    {
        //左下角的子表中数据项为左上角子表对应项加 2^(i-1)
        for(ii = 0; ii<subTableLen; ++ii)
            for(jj=0; jj<subTableLen; ++jj)
                tables[ii+subTableLen][jj] = tables[ii][jj] +subTableLen;

        //右上角的子表等于左下角子表
        for(ii = 0; ii<subTableLen; ++ii)
            for(jj=0; jj<subTableLen; ++jj)
                tables[ii][jj+subTableLen] = tables[ii+subTableLen][jj];

        //右下角的子表等于左上角子表
        for(ii = 0; ii<subTableLen; ++ii)
```

```
            for(jj=0; jj<subTableLen; ++jj)
                tables[ii+subTableLen][jj+subTableLen] = tables[ii][jj];

        subTableLen*=2;
    }

    //print results
    for(i=0; i<n; ++i)
    {
        for(j=0; j<n; ++j)
        {
            cout<<setw(3)<<tables[i][j];
        }
        cout<<endl;
        delete[] tables[i];
    }
    delete[] tables;
}
int main()
{
    int k;
    cout<<"Input k to get 2^k players' match agenda: ";
    cin>>k;
    if(k<0)
        return 0;
    Agenda(k);
}
```


上述解法已经能处理 2^k 个选手的循环赛比赛日程表，但如果选手数目不是 2 的指数，而是任意个数 n，则意味比赛日程可能会有轮空现象。可以将上述代码修改如下：

```
void AgendaAnyN(int n)
{
    int i, j, subTableLen, ii, jj, k=0, alignN=1;
    //找到一个 k，使得 2^k>=n
    while(alignN<n)
    {
        alignN <<=1;
        ++k;
    }
    //只有一位选手时
    if(k==0)
    {
        k=1;
        alignN = 2;
```

```
    }

    int **tables = new int*[alignN];
    for(i=0; i<alignN; ++i)
        tables[i] = new int[alignN];
    //初始化两人比赛问题的日程表
    tables[0][0] = 1;
    tables[0][1] = 2;
    tables[1][0] = 2;
    tables[1][1] = 1;
    subTableLen = 2;
    for(i=2; i<=k; ++i)
    {
        //左下角的子表中数据项为左上角子表对应项加 2^(i-1)
        for(ii = 0; ii<subTableLen; ++ii)
            for(jj=0; jj<subTableLen; ++jj)
                tables[ii+subTableLen][jj] = tables[ii][jj] +subTableLen;

        //右上角的子表等于左下角子表
        for(ii = 0; ii<subTableLen; ++ii)
            for(jj=0; jj<subTableLen; ++jj)
                tables[ii][jj+subTableLen] = tables[ii+subTableLen][jj];

        //右下角的子表等于左上角子表
        for(ii = 0; ii<subTableLen; ++ii)
            for(jj=0; jj<subTableLen; ++jj)
                tables[ii+subTableLen][jj+subTableLen] = tables[ii][jj];

        subTableLen*=2;
    }

    //打印结果
    for(i=0; i<n; ++i)
    {
        for(j=0; j<alignN; ++j)
        {
            if(tables[i][j]>n)
                cout<<setw(3)<<"/"; //表示轮空
            else
                cout<<setw(3)<<tables[i][j];
        }
        cout<<endl;
        delete[] tables[i];
    }
```

```
    delete[] tables;
}
```

例如，输入 5 个选手时，比赛日程表如下：

```
1 2 3 4 5 / / /
2 1 4 3 / 5 / /
3 4 1 2 / / 5 /
4 3 2 1 / / / 5
5 / / / 1 2 3 4
```

5.7 棋盘覆盖

在一个 $2^k \times 2^k$ 个方格组成的棋盘中，若恰有一个方格与其他方格不同，则称该方格为特殊方格，称该棋盘为特殊棋盘。特殊方格在棋盘中出现的位置有 $2^k \times 2^k$ 种情形。特殊棋盘如图 5-2 所示，是当 $k=2$ 时 16 个特殊棋盘中的一个。在棋盘覆盖问题中，要求用如图 5-3 所示的四种不同形状的 L 型骨牌覆盖给定棋盘上除特殊方格以外的所有方格，且任何两个 L 型骨牌不得重叠覆盖。在任何一个 $2^k \times 2^k$ 的棋盘覆盖中，用到的 L 型骨牌个数显然为 $(2^k \times 2^k-1)/3$。

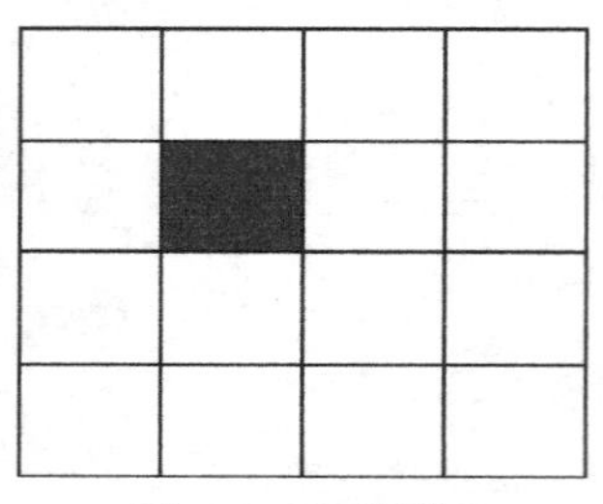

图 5-2 特殊棋盘

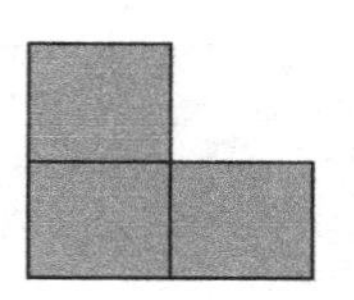
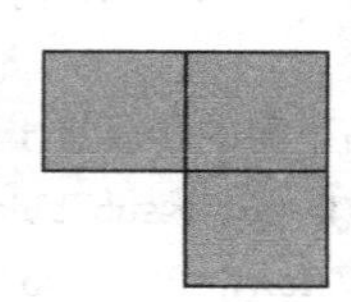
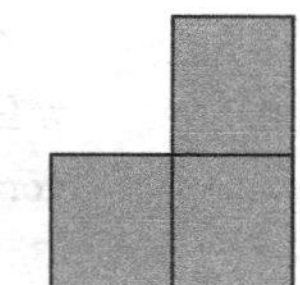

图 5-3 四种 L 型骨牌

这里考虑使用分治方法，将 $2^k \times 2^k$ 棋盘覆盖问题转换成 4 个 $2^{k-1} \times 2^{k-1}$ 棋盘覆盖问题。分治的策略在于如何划分棋盘，并且使每个子棋盘均包含一个特殊方格，从而将原问题分解为规模较小的棋盘覆盖问题。根据上述内容，归纳如下分治策略。

（1）当 $k>0$ 时，将 $2^k \times 2^k$ 的棋盘划分为四个 $2^{k-1} \times 2^{k-1}$ 子棋盘，分割成四块子棋盘如图 5-4 所示。

（2）此时只有一个子棋盘中有一个特殊方格，其余三个子棋盘中没有特殊方格。所以需要用一个 L 型骨牌覆盖这三个较小棋盘的交汇处。从而将原问题转化为四个较小规模的棋盘覆盖问题，如图 5-5 所示。

（3）递归该分治划分方法，直至将棋盘分割为 1×1 的子棋盘。

例如，输入 $k=2$，特殊方格的位置是（1,1），则一个可行的覆盖结果如下：

```
2 2 3 3
2 0 1 3
4 1 1 5
4 4 5 5
```

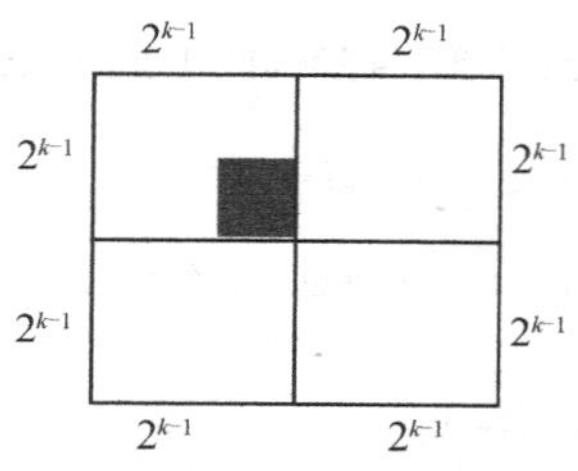

图 5-4　分割成四块子棋盘

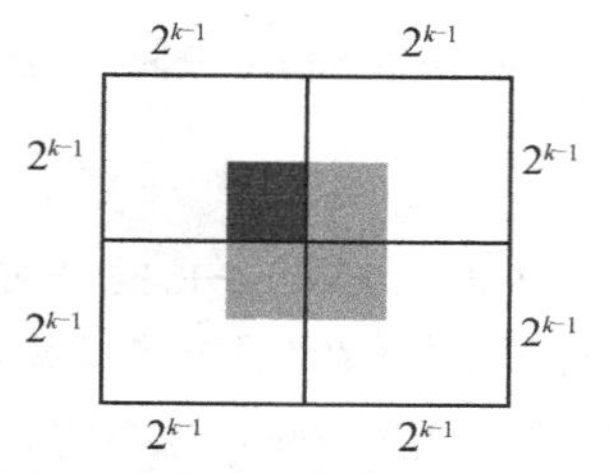

图 5-5　转化为四个较小规模的棋盘覆盖问题

覆盖结果是一个二维数组，数组中每个元素的值是 L 型骨牌铺设时的编号，编号从 1 开始，编号为 0 的元素表示输入特殊方格。代码如下：

```
#include <iostream>
#include <iomanip>

using namespace std;

//L 型骨牌铺设时的编号
int lastLNum = 0;

//求解棋盘覆盖的递归函数
//chessboard 是棋盘二维数组，元素值代表铺设 L 型骨牌的编号，编号从 1 开始
//leftX, leftY，表示当前待求解的子棋盘的左上角位置
//width 表示当前待求解的子棋盘的宽度，该值一定是 2 的指数
//specialGridX，specialGridY 表示当前待求解的子棋盘的特殊方格的位置
void SolveChessBoard(int**chessboard, int leftX, int leftY, int width, int specialGridX, int specialGridY)
{
    if(width= =1)
        return;
    int subWidth = width/2;
    int currentLNum = ++lastLNum;
    //特殊方格不在左上角子棋盘上
    if(!(leftX+subWidth>specialGridX && leftY+subWidth>specialGridY))
    {
        //铺设新的 L 型骨牌，添加了新的特殊方格
        chessboard[leftX+subWidth-1][leftY+subWidth-1] = currentLNum;
        SolveChessBoard(chessboard, leftX, leftY, subWidth, leftX+subWidth-1, leftY+subWidth-1);
    }
    else
        SolveChessBoard(chessboard, leftX, leftY, subWidth, specialGridX, specialGridY);
    //特殊方格不在左下角子棋盘上
    if(!(leftX+subWidth>specialGridX && leftY+subWidth<=specialGridY))
    {
        //铺设新的 L 型骨牌，添加了新的特殊方格
        chessboard[leftX+subWidth-1][leftY+subWidth] = currentLNum;
```

```
        SolveChessBoard(chessboard, leftX, leftY+subWidth, subWidth, leftX+subWidth-1, leftY+subWidth);
    }
    else
        SolveChessBoard(chessboard, leftX, leftY+subWidth, subWidth, specialGridX, specialGridY);
    //特殊方格不在右上角子棋盘上
    if(!(leftX+subWidth<=specialGridX && leftY+subWidth>specialGridY))
    {
        //铺设新的L型骨牌，添加了新的特殊方格
        chessboard[leftX+subWidth][leftY+subWidth-1] = currentLNum;
        SolveChessBoard(chessboard, leftX+subWidth, leftY, subWidth, leftX+subWidth, leftY+subWidth-1);
    }
    else
        SolveChessBoard(chessboard, leftX+subWidth, leftY, subWidth, specialGridX, specialGridY);
    //特殊方格不在右下角子棋盘上
    if(!(leftX+subWidth<=specialGridX&& leftY+subWidth<=specialGridY))
    {
        //铺设新的L型骨牌，添加了新的特殊方格
        chessboard[leftX+subWidth][leftY+subWidth] = currentLNum;
        SolveChessBoard(chessboard, leftX+subWidth, leftY+subWidth, subWidth, leftX+subWidth, leftY+
subWidth);
    }
    else
        SolveChessBoard(chessboard, leftX+subWidth, leftY+subWidth, subWidth, specialGridX, specialGridY);
}

//求解棋盘覆盖，k表示棋盘宽度的指数
//specialGridX，specialGridY表示输入的棋盘上特殊方格的位置
void CoverChessBoard(int k, int specialGridX, int specialGridY)
{
    int i, j;
    int n = 1<<k;
    //分配一个二维数组
    int **chessboard = new int*[n];
    for(i=0; i<n; ++i)
        chessboard[i] = new int[n];
    //设置特殊方格所在的棋盘值为0
    chessboard[specialGridX][specialGridY] = 0;
    lastLNum = 0;
    SolveChessBoard(chessboard, 0, 0, n, specialGridX, specialGridY);
    for(i=0; i<n; ++i)
    {
        for(j=0; j<n; ++j)
            cout<<setw(3)<<chessboard[i][j];
        cout<<endl;
```

```
        }
        for(i=0; i<n; ++i)
            delete[] chessboard[i];
        delete[] chessboard;
}

int main()
{
        int k, specialGridX, specialGridY;
        cout<<"Input k to set up 2^k*2^k chess board: ";
        cin>>k;
        cout<<"Input one special grid x and y:";
        cin>>specialGridX >> specialGridY;

        CoverChessBoard(k, specialGridX, specialGridY);
        return 0;
}
```

第 6 章　集合与映射

与 C 语言相比，C++语言在算法设计时有一个天然优势：STL。STL 中除已有的大量算法函数外，还提供了极其有用的模板容器。前面章节介绍了 std::vector, std::list, std::deque 容器在解决实际问题时的用法。本章将讨论集合容器 std::set、映射容器 std::map 及哈希映射。

6.1　集合

在数学理论中，集合中所有元素的值都是唯一的，STL 的 std::set 就是这样一种集合容器。

6.1.1　集合 std::set

集合（std::set）是一个容器，该容器中所包含的元素的值都是唯一的。除了元素的唯一性外，容器内的元素还按照一定的顺序排列。std::set 通常采用二分搜索树实现。

使用 std::set 需要引入头文件<set>。如果不在乎元素的内部顺序性，还可以使用 std::unordered_set 容器。std::unordered_set 是一种无序的集合容器，其内部元素不是有序的，但该容器创建和访问速度都优于 std::set。如果希望元素的键值可以多次出现，那么可以使用多集容器 std::multiset，该容器中同一值可以出现多次。

std::set 集合创建与遍历的样例程序如下：

```
#include <iostream>
#include<set>
using namespace std;
int main()
{
    int data[]={12, 34, 10, 98, 3, 10, 12};
    std::set<int> dataSet;
    for(size_t i=0; i<sizeof(data)/sizeof(data[0]); ++i)
    {
            //把数据插入集合，数据自动排序
            dataSet.insert(data[i]);
    }
    //输出已排序的集合元素
```

```
    for(std::set<int>::iterator it=dataSet.begin(); it!=dataSet.end(); ++it)
            std::cout << *it << " ";
}
```

输出：

```
3  10  12  34  98
```

如果想让集合中元素降序排列，则代码修改如下：

```
#include <iostream>
#include<set>
using namespace std;
int main()
{
    int data[]={12, 34, 10, 98, 3, 10, 12};
    std::set<int, std::greater<int> > dataSet;
    for(size_t i=0; i<sizeof(data)/sizeof(data[0]); ++i)
    {
            //把数据插入集合，数据自动排序
            dataSet.insert(data[i]);
    }

    //输出已排序的集合元素
    for(std::set<int, std::greater<int> >::iterator it=dataSet.begin(); it!=dataSet.end(); ++it)
            std::cout << *it << " ";
}
```

输出：

```
98  34  12  10  3
```

std::set 集合查找与删除的示例程序如下：

```
std::set<int>::iterator it=dataSet.find(98);
if(it!=dataSet.end())
{
   cout<<" Find it" ;
   dataSet.erase(it); //删除
}
else
   cout<<" Not in dataSet" ;
```

6.1.2 集合求交 set_intersection

STL 提供了集合的交、并、差集函数。其中集合交集的函数是 std::set_intersection，函数声明如下：

```
template <class InputIterator1, class InputIterator2, class OutputIterator>
```

```
OutputIterator set_intersection (InputIterator1 first1, InputIterator1 last1,InputIterator2 first2, InputIterator2 last2,OutputIterator result);
```

前两个参数是第一个集合的范围，第三和第四个参数是第二个集合的范围，最后一个参数是输出结果集合的起始位置。在使用 std::set_intersection 之前，请务必确保输入的两个集合都是已排序的集合。函数的返回值是一个迭代器，它指向了输出结果域的下一个元素位置。再次强调，该返回值不一定指向输出结果集合的 end 位置，而是指向求交后域(range)的最后一个迭代器位置。例如，集合 A 和集合 B，共有相同的元素 4 个，那么该返回值迭代器位置等于 result+=4。

std::set_intersection 的内部实现类似于如下代码：

```
template <class InputIterator1, class InputIterator2, class OutputIterator>
  OutputIterator set_intersection (InputIterator1 first1, InputIterator1 last1,InputIterator2 first2, InputIterator2 last2,OutputIterator result)
{
  while (first1!=last1 && first2!=last2)
  {
    if (*first1<*first2) ++first1;
    else if (*first2<*first1) ++first2;
    else {
      *result = *first1;
      ++result; ++first1; ++first2;
    }
  }
  return result;
}
```


std::set_intersection 的另一个重载函数声明如下：

```
template <class InputIterator1, class InputIterator2,   class OutputIterator, class Compare>
  OutputIterator set_intersection (InputIterator1 first1, InputIterator1 last1,InputIterator2 first2, InputIterator2 last2, OutputIterator result, Compare comp);
```

该重载函数多了一个比较函数作为最后一个参数。

std::set_intersection 示例代码如下：

```
#include <iostream>       // std::cout
#include <algorithm>      // std::set_intersection, std::sort
#include <vector>         // std::vector
int main ()
{
    int first[] = {1,5,9,3,7};
    int second[] = {5,4,3,30,20,10};
    int size1 = sizeof(first)/sizeof(first[0]);
    int size2 = sizeof(second)/sizeof(second[0]);
    std::vector<int> v;
    v.resize(size1+size2);// 0  0  0  0  0  0  0  0  0  0  0
    std::vector<int>::iterator it;
```

```
        std::sort (first,first+size1);          //   1 3 5 7 9
        std::sort (second,second+size2);     //3 4 5 10 20 30
        it=std::set_intersection (first, first+size1, second, second+size2, v.begin()); // 3 5 0 0 0 0 0 0 0 0
        v.resize(it-v.begin());                                  //3 5
        std::cout << "The intersection has " << (v.size()) << " elements:\n";
        for (it=v.begin(); it!=v.end(); ++it)
            std::cout << ' ' << *it;
        std::cout << '\n';
        return 0;
}
```

输出：

```
The intersection has 2 elements:
3   5
```

在使用 std::set_intersection 之前，程序对 first 和 second 数组分别按升序进行了排序，并且创建一个 std::vector 对象 v 用来存储交集计算的结果。需要特别注意的是 v 中必须有足够的空间来存储交集结果，所以这里调用了 v.resize(size1+size2)预留足够空间。另外，交集计算完成后，还需要把 v 中多余的空间删掉，即调用 v.resize(it-v.begin())。之前我们强调返回值 it 指向求交集合域最后一个迭代器位置。所以可以用 it-v.begin()计算求教集合实际需要的大小。

有没有更简单的方法？答案是使用 std::back_insert。使用 std::back_insert 时，根本不需要为结果集合预留任何空间，也不需要在计算完成后删除多余的空间。读者可以将上述 std::set_intersection 语句替换成如下语句，需要引入头文件<iterator>，代码如下：

```
v.clear();
std::set_intersection (first, first+size1, second, second+size2, std::back_inserter(v));
```

最后一个疑问是如果 first 和 second 都是降序排列的，std::set_intersection 该如何调用？示例代码如下：

```
    std::sort (first,first+size1, std::greater<int>());
    std::sort (second,second+size2, std::greater<int>());
```

那么此时需要调用 set_intersection 的另一个重载函数，代码如下：

```
    it=std::set_intersection (first, first+size1, second, second+size2, v.begin(), std::greater<int>());
```

或者，代码如下：

```
v.clear();
std::set_intersection (first, first+size1, second, second+size2, std::back_inserter(v), std::greater<int>());
```

6.1.3 集合求并 set_union

集合并集函数 std::set_union 示例代码如下：

```
#include <iostream>       // std::cout
#include <algorithm>      // std::set_union, std::sort
```

```
#include <vector>            // std::vector
#include <iterator>          // std::back_inserter
int main ()
{
    int first[] = {1,5,9,3,7};
    int second[] = {5,4,3,30,20,10};
    int size1 = sizeof(first)/sizeof(first[0]);
    int size2 = sizeof(second)/sizeof(second[0]);
    std::vector<int> v;
    std::vector<int>::const_iterator it;
    std::sort (first,first+size1, std::greater<int>());
    std::sort (second,second+size2, std::greater<int>());
    std::set_union (first, first+size1, second, second+size2, std::back_inserter(v), std::greater<int>());
    std::cout << "The union has " << (v.size()) << " elements:\n";
    for (it=v.begin(); it!=v.end(); ++it)
        std::cout << ' ' << *it;
    std::cout << '\n';
    return 0;
}
```

输出：


```
The union has 9 elements:
30  20  10  9  7  5  4  3  1
```

6.1.4 集合求差 set_difference

集合求差集函数 std::set_difference 示例代码如下：

```
#include <iostream>          // std::cout
#include <algorithm>         // std::set_difference, std::sort
#include <vector>            // std::vector
#include <iterator>          // std::back_inserter
int main ()
{
    int first[] = {1,5,9,3,7};
    int second[] = {5,4,3,30,20,10};
    int size1 = sizeof(first)/sizeof(first[0]);
    int size2 = sizeof(second)/sizeof(second[0]);
    std::vector<int> v;
    std::vector<int>::const_iterator it;
    std::sort (first,first+size1, std::greater<int>());
    std::sort (second,second+size2, std::greater<int>());
    std::set_difference (first, first+size1, second, second+size2, std::back_inserter(v), std::greater<int>());
    std::cout << "first - second difference has " << (v.size()) << " elements:\n";
    for (it=v.begin(); it!=v.end(); ++it)
```

```
        std::cout << ' ' << *it;
    std::cout << '\n';
    v.clear();
    std::set_difference (second, second+size2, first, first+size1,  std::back_inserter(v), std::greater<int>());
    std::cout << "second - first difference has " << (v.size()) << " elements:\n";
    for (it=v.begin(); it!=v.end(); ++it)
        std::cout << ' ' << *it;
    std::cout << '\n';
    return 0;
}
```

输出：

```
first - second difference has 3 elements:
9  7  1
second - first difference has 4 elements:
30  20  10  4
```

6.1.5 综合编程实例：集合相似度

给定两个整数集合，集合相似度定义为：Nc/Nt*100%。其中 Nc 是两个集合都有的不相等整数的个数，Nt 是两个集合一共有的不相等整数的个数。要求计算任意一对给定集合的相似度。

在第一行输入一个正整数 N（<=50），表示集合的个数。随后 N 行，每行对应一个集合。每个集合首先给出一个正整数 M（$<=10^4$），表示集合中元素的个数；然后输入 M 个整数，每个整数的取值范围为$[0，10^9]$。之后一行给出一个正整数 K（<=2000），随后 K 行，每行对应一对需要计算相似度的集合的编号（集合从 1 到 N 编号）。数字间以空格分隔。对每对需要计算的集合，在一行中输出它们的相似度，保留小数点后 2 位的百分比数字。

输入示例：

```
3
3 99 87 101
4 87 101 5 87
7 99 101 18 5 135 18 99
2
1 2
1 3
```

输出示例：

```
50.00%
33.33%
```

这是团体程序设计天梯赛——练习集 L2-005 题目（作者陈越，链接地址: https://pintia.cn/problem-sets/994805046380707840/problems/994805070149828608）。只要读者掌握了 STL 集合

的交与并的用法，该题自然迎刃而解，代码如下：

```
#include <iostream>
#include <string>
#include <set>
#include <vector>
#include <algorithm>
#include <iomanip>
using namespace std;

#define MAX_N 51
int main()
{
    int n, i, m, j, v, K, id1, id2;
    std::set<int> rawDatas[MAX_N];
    cin>>n;
    for(i=0; i<n; ++i)
    {
        cin>>m;
        for(j=0; j<m; ++j)
        {
            cin>>v;
            rawDatas[i].insert(v);
        }
    }
    cin>>K;
    for(i=0; i<K; ++i)
    {
        cin>>id1>>id2;
        id1-=1;
        id2-=1;
        std::vector<int> interV, unionV;
        //计算差集
    std::set_intersection(rawDatas[id1].begin(), rawDatas[id1].end(), rawDatas[id2].begin(), rawDatas[id2].end(), std::bac
k_inserter(interV));
        //计算并集
    std::set_union(rawDatas[id1].begin(), rawDatas[id1].end(), rawDatas[id2].begin(), rawDatas[id2].end(), std::
back_inserter(unionV));
        cout<<std::fixed<<setprecision(2)<<interV.size()/(unionV.size()+0.0)*100<<"%"<<endl;
    }
}
```

6.2 映射

在 STL 中，映射(std::map)是一种关联性容器，该容器存储的是键值对 (key value pair)。在该容器对象建立过程中，键都是已排序的。键的数据类型可以和值的类型不同。std::map 的搜索、删除及插入操作速度很快，因为 std::map 普遍采用二分搜索树实现。

6.2.1 std::map 基本用法

使用 std::map 需要引入头文件<map>。一段 std::map 的简单代码如下：

```
#include <iostream>
#include <map>
#include <string>
using namespace std;
int main()
{
    map<string, float> nameMapGPA;
    nameMapGPA.insert(std::make_pair("Bob", 3.2));
    nameMapGPA.insert(std::make_pair("Adam", 3.5));
    nameMapGPA.insert(std::make_pair("Stanley", 2.1));
    nameMapGPA.insert(std::make_pair("Tim", 2.8));

    map<string, float>::iterator iter;
    for(iter = nameMapGPA.begin(); iter != nameMapGPA.end(); iter++)
        cout<<iter->first<<' '<<iter->second<<endl;
    return 0;
}
```

输出：

```
Adam 3.5
Bob 3.2
Stanley 2.1
Tim 2.8
```

上述代码中，map 容器的键值对类型是<string, float>，也就是使用名字作为键，成绩为值。插入键值对时需要使用 std::make_pair 函数先创建一个键值对，然后逐一调用 map 的成员函数 insert 插入键值对。

如果想让成绩作为键，名字为值，并且遍历的时候希望将成绩高的排在前面，则代码可修改如下：

```
#include <iostream>
#include <map>
#include <string>
```

```
using namespace std;

int main()
{
    map<float, string, std::greater<float> > GPAMapName;
    GPAMapName.insert(std::make_pair(3.2, "Bob"));
    GPAMapName.insert(std::make_pair(3.5, "Adam"));
    GPAMapName.insert(std::make_pair(2.1, "Stanley"));
    GPAMapName.insert(std::make_pair(2.8, "Tim"));

    map<float, string, std::greater<float> >::iterator iter;
    for(iter = GPAMapName.begin(); iter != GPAMapName.end(); iter++)
        cout<<iter->first<<' '<<iter->second<<endl;
}
```

输出：

```
3.5 Adam
3.2 Bob
2.8 Tim
2.1 Stanley
```

6.2.2 std::multimap 基本用法

上一节给出的两段代码都有一个问题：输入数据中的键不能相同。这意味第一段代码中不能有同名的学生，第二段代码中不能有 GPA 相同的学生。这种限制条件显然过于理想化，碰到这种情况就要使用 std::multimap 容器，该类型容器允许相同的键出现。此时，可以将第 6.2.1 节中第一段代码修改成如下鲁棒性更强的代码，代码如下：

```
#include <iostream>
#include <map>
#include <string>
using namespace std;
int main()
{
    multimap<string, float> nameMapGPA;
    nameMapGPA.insert(std::make_pair("Bob", 3.2));
    nameMapGPA.insert(std::make_pair("Adam", 3.5));
    nameMapGPA.insert(std::make_pair("Adam", 1.5));
    nameMapGPA.insert(std::make_pair("Stanley", 2.1));
    nameMapGPA.insert(std::make_pair("Tim", 2.8));
    nameMapGPA.insert(std::make_pair("Stanley", 3.2));

    multimap<string, float>::iterator iter;
    for(iter = nameMapGPA.begin(); iter != nameMapGPA.end(); iter++)
        cout<<iter->first<<' '<<iter->second<<endl;
    return 0;
```

```
}
```

输出：

```
Adam 3.5
Adam 1.5
Bob 3.2
Stanley 2.1
Stanley 3.2
Tim 2.8
```

将第 6.2.1 节中第二段代码修改成如下鲁棒性更强的代码，代码如下：

```
#include <iostream>
#include <map>
#include <string>
using namespace std;

int main()
{
    multimap<float, string, std::greater<float> > GPAMapName;
    GPAMapName.insert(std::make_pair(1.5, "Adam"));
    GPAMapName.insert(std::make_pair(3.2, "Bob"));
    GPAMapName.insert(std::make_pair(3.5, "Adam"));
    GPAMapName.insert(std::make_pair(2.1, "Stanley"));
    GPAMapName.insert(std::make_pair(2.8, "Tim"));
    GPAMapName.insert(std::make_pair(3.2, "Stanley"));

    multimap<float, string>::iterator iter;
    for(iter = GPAMapName.begin(); iter != GPAMapName.end(); iter++)
        cout<<iter->first<<' '<<iter->second<<endl;
}
```

输出：

```
3.5 Adam
3.2 Bob
3.2 Stanley
2.8 Tim
2.1 Stanley
1.5 Adam
```

6.3 哈希映射

哈希表（Hash Table）也叫散列表，是根据键值对（key valuepair）而直接进行访问的数据结构。哈希表通过把关键键值映射到哈希表中的一个位置来访问记录，从而加快查找的速度。这个映射函数就叫作哈希函数或散列函数，而存放记录的数组叫作哈希表。

哈希表存储的内容是键值对，其查找的时间复杂度与元素数量多少无关，哈希表在查找元素时是通过计算哈希码值来定位元素的位置，因此哈希表查找的时间复杂度为 O(1)。

6.3.1 常用的哈希函数

以数据中每个元素的关键字 K 为自变量，通过散列函数 H（K）计算函数值，以该函数值作为一块连续存储空间的单元地址，将该元素存储到函数值对应的单元中。

1. 除留余数法

取关键字被某个不大于散列表表长 L 的数 P 除后所得的余数为散列地址，即 H（Key）= Key MOD P，P<=L。对 P 的选择很重要，若 P 选得不好，则很容易产生冲突。一般直接将 P 取值为表的长度 L，L 一般定义为质数。

2. 直接寻址法

取关键字或者关键字的某个线性函数值作为哈希地址，即 H（Key）=Key 或者 H（Key）= a*Key+b（a,b 为整数）。如果 H（Key）的哈希地址上已经有值，那么就往下一个位置查找，直到找到 H(Key)的位置没有值就把元素放进去。

6.3.2 哈希冲突的处理方法

1. 线性探测法

地址增量 $d = 1, 2, \ldots, L-1$。该方法依次探测下一个地址，直到找到空闲的空间后插入，若整个空间都找不到空余的空间，则产生溢出。

2. 平方探测法

地址增量 $d = 1^2, -1^2, 2^2, -2^2, \ldots$,（L/2）^{2}, –（L/2）^{2}。该方法依次探测下一个平方地址。该方法能有效避免“聚集”现象，但是不能够探测到哈希表上所有的存储单元，但是至少能够探测到一半。

3. 链地址法

将所有具有相同哈希地址而不同关键字的数据元素连接到同一个单链表中。所以该方法能彻底解决溢出问题。

6.3.3 综合编程实例

1. 哈希冲突的处理方法

给定一系列整型关键字和素数 P,用除留余数法定义的散列函数将关键字映射到长度为 P 的散列表中。用线性探测法解决冲突。

在第一行输入两个正整数 N（≤1000）和 P（≥N 的最小素数），分别为待插入的关键字总数，以及散列表的长度。第二行给出 N 个整型关键字。数字间以空格分隔。

输出格式：

在一行内输出每个整型关键字在散列表中的位置。数字间以空格分隔，但行末尾不得有多余空格。

输入样例：

```
4 5
24 15 61 88
```

输出样例：

```
4 0 1 3
```

这道题出自PTA题库（作者：DS课程组，单位：浙江大学），题目地址：https://pintia.cn/problem-sets/15/problems/889。

题目要求使用线性探测法解决冲突，新探测位置 $S_{new}=(S_{raw}+i)\%P$, i=1,2,3,...,L−1。示例代码如下：

```
#include <iostream>
#include <cmath>

#define MAX_P 1009
using namespace std;
//找到>=N 的最小素数，这里用来计算 MAX_P，即 1009
int MinPrime(int N)
{
    int P=N, i, j;
    while(true)
    {
        j = sqrt(P);
        for(i=2; i<=j; ++i)
        {
            if(P%i==0)
                break;
        }
        if(i==j+1)
            return P;
        ++P;
    }
    return 0;
}
//全局哈希表数组，数组中所有元素初始为 0，表示没有被占据
int hashTable[MAX_P];
int main()
{
    int N, P, i, pos, value;
```

```
    bool bFirst=true;
    cin>>N>>P;
    for(i=0; i<N; ++i)
    {
        cin>>value;
        pos = value%P;
        //如果映射位置没有被占据，或者被占据的值相同
        if(hashTable[pos]==0||hashTable[pos]==value)
        {
            hashTable[pos] = value;
            if(!bFirst)
                cout<<" ";
            cout<<pos;
        }
        else
        {
           //有冲突
            while(true)
            {
                //采用线性探测法
                pos = (pos+1)%P;
                //如果映射位置没有被占据，或者被占据的值相同
                if(hashTable[pos]==0||hashTable[pos]==value)
                {
                    hashTable[pos] = value;
                    if(!bFirst)
                        cout<<" ";
                    cout<<pos;
                    break;
                }
            }
        }
        bFirst=false;
    }
    return 0;
}
```

2. 字符串关键字的散列映射

给定一系列由大写英文字母组成的字符串关键字和素数 P，用移位法定义的散列函数 H（Key）将关键字 Key 中的最后 3 个字符映射为整数，每个字符占 5 位；再用除留余数法将整数映射到长度为 P 的散列表中。例如，将字符串 AZDEG 插入长度为 1009 的散列表中，我们首先将 26 个大写英文字母顺序映射到整数 0～25；再通过移位将其映射为 $3\times32^2+4\times32^1+6=3206$；然后根据表长得到该字符串的散列映射位置。

发生冲突时请用平方探测法解决。

输入格式：

在第一行输入两个正整数 N（≤500）和 P（≥$2N$ 的最小素数），分别为待插入的关键字总数，以及散列表的长度。第二行给出 N 个字符串关键字，每个长度不超过 8 位，其间以空格分隔。

输出格式：

在一行内输出每个字符串关键字在散列表中的位置。数字间以空格分隔，但行末尾不得有多余空格。

输入样例 1：

```
4 11
HELLO ANNK ZOE LOLI
```

输出样例 1：

```
3 10 4 0
```

输入样例 2：

```
6 11
LLO ANNA NNK ZOJ INNK AAA
```

输出样例 2：

```
3 0 10 9 6 1
```

这道题同样出自 PTA 题库（作者：DS 课程组，单位：浙江大学），题目地址：https://pintia.cn/problem-sets/15/problems/890。

既然冲突时要求使用平方探测法，那么新探测位置 S_{new}=（$S_{raw}\pm i^2$）%P，i=1,2,3,...,L−1。需要注意，如果 $S_{raw}-i^2$ 为负数时，需要将其加 $K*P$，使其变为正数。代码如下：

```
#include <iostream>
#include <cmath>
using namespace std;
#define MAX_P 1009
std::string hashTabel[MAX_P];
//找到>=2*N 的最小素数
int MinPrime(int N)
{
    int P = 2*N;
    int i, t;
    while(true)
    {
        t=sqrt(P);
        for(i=2; i<=t; ++i)
        {
            if(P%i==0)
                break;
```

```
            }
            if(i>t)
                break;
            ++P;
        }
        return P;
}
int main()
{
        int N, P, i, j, k, len, sum, pow32;

        std::string str;
        cin>>N>>P;
        for(k=0; k<N; ++k)
        {
            cin>>str;
            len = str.length();
            //计算哈希值
            sum = 0;
            pow32 = 1;
            j = std::max(len-3, 0);
            for(i=len-1; i>=j; --i)
            {
                sum+=(str[i]-'A')*pow32;
                pow32*=32;
            }
            i = sum%P;
            if(hashTabel[i].empty()||hashTabel[i]==str)
            {
                hashTabel[i] = str;
                if(k!=0)
                    cout<<" ";
                cout<<i;
            }
            else
            {
                for(j=1; ;++j)
                {
                    //先正向进行平方探测，再反向进行平方探测
                    if(hashTabel[(i+j*j)%P].empty()||hashTabel[(i+j*j)%P]==str)
                    {
                        i = (i+j*j)%P;
                        hashTabel[i] = str;
                        if(k!=0)
```

```
                        cout<<" ";
                    cout<<i;
                    break;
                }
                else
        if(hashTabel[(i-j*j+1000*P)%P].empty()||hashTabel[(i-j*j+1000*P)%P]==str)
                {
                    i = (i-j*j+1000*P)%P;
                    hashTabel[i] = str;
                    if(k!=0)
                        cout<<" ";
                    cout<<i;
                    break;
                }
            }
        }
    }
    return 0;
}
```

第 7 章　Win32 GUI 编程基础

在程序设计的初级阶段，使用 Console 程序即控制台程序能解决很多基本问题。但控制台程序的交互性不足，对图像和图形数据的可视化较为困难。作为综合程序设计，需要引入 GUI 程序设计作为补充。GUI 是 Graphics User Interface 的简写，即图形用户接口。

本章将使用 CodeBlocks 和 Visual Studio 2015 Community 两个免费的集成开发环境（Integrated Development Environment，IDE）创建 Win32 GUI 程序。

CodeBlocks 是一款免费的 IDE，该 IDE 支持 C、C++、Fortran 等语言的开发。读者可以从网站（http://www.codeblocks.org/downloads/26）下载该软件。如果读者不熟悉手动配置 gcc 和 g++的编译器或调试器，那么下载 CodeBlocks 时，需要选择自带 mingw 的版本下载，如 codeblocks-17.12mingw-setup.exe 的版本。当版本更新后数字 17.12 也会对应升级。

Visual Studio 2015 Community 版本是微软公司免费的 IDE，但相比于 CodeBlocks，Visual Studio 2015 Community 大约需要 6GB 安装空间，非常庞大。另外，它不支持离线下载安装，必须在线安装。读者可以从网址（https://www.visualstudio.com/zh-hans/downloads/）安装该 IDE。

7.1　Win32 GUI

7.1.1　CodeBlocks 第一个 Win32 教程

安装 CodeBlocks 后，打开 CodeBlocks，选择“File”→“New”→“Project...”选项（如图 7-1 所示），然后在新建模板对话框中选择“Win32 GUI project”图标（如图 7-2 所示），接着选择“Frame based”选项（如图 7-3 所示），单击“Next”按钮，将项目标题命名成“FirstTutorial”（如图 7-4 所示）。需要指出的是，CodeBlocks 的项目最好不要放在中文路径下，因为 GDB 调试器在中文路径下通常无法正常使用。另外项目名称最好使用英文命名，单词之间最好不要加空格。

上述操作完成后，CodeBlocks 就会自动创建一些代码（如图 7-5 所示）。选择“Build”→“Build and run”或者按快捷键 F9 直接对该工程进行编译并运行。运行结果如图 7-6 所示。

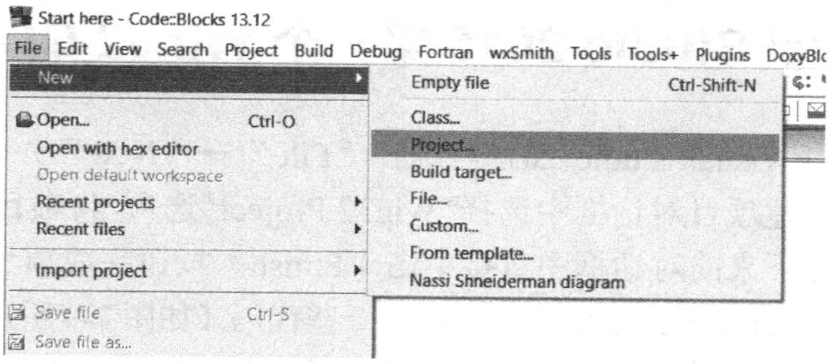

图 7-1　选择“File”→“New”→“Project”选项

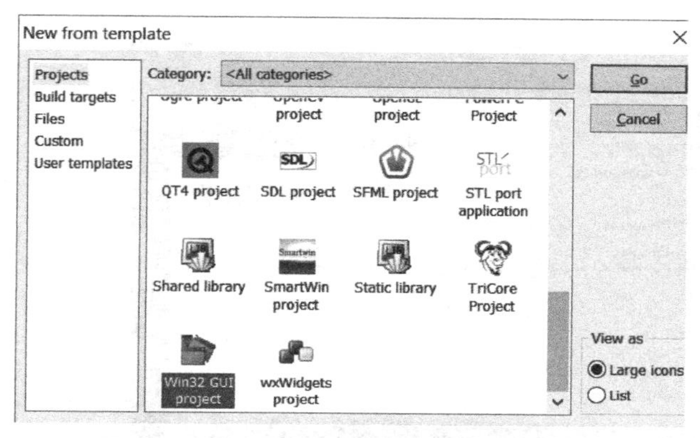

图 7-2　选择“Win32 GUI project”图标

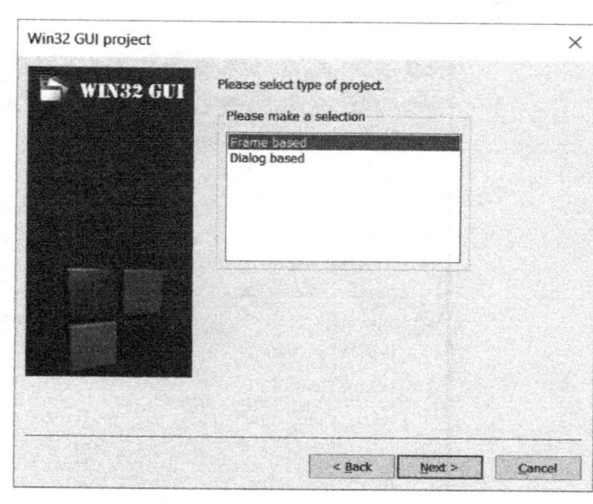

图 7-3　选择“Frame based”选项

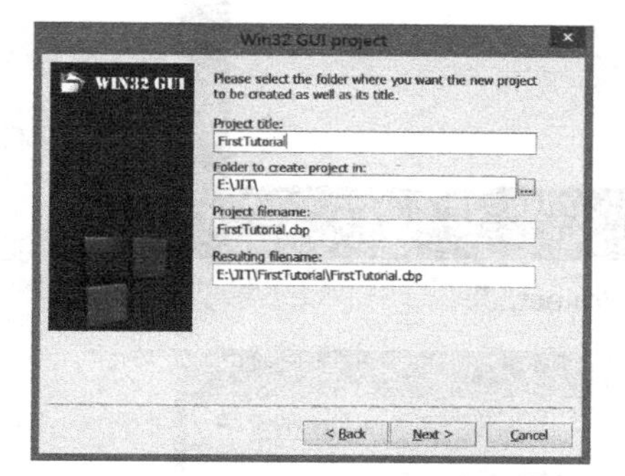

图 7-4　命名项目名称

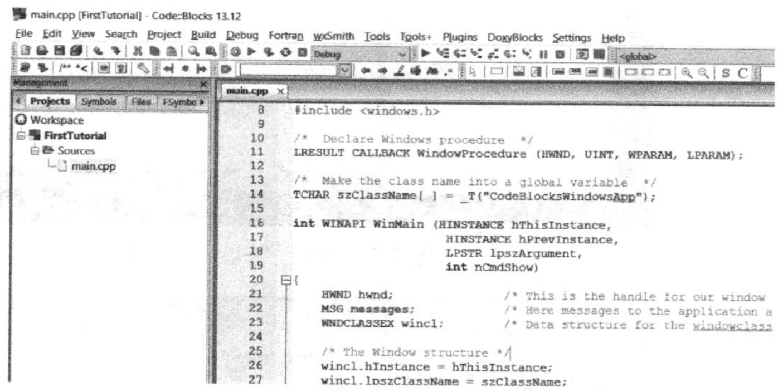

图 7-5　自动生成的代码

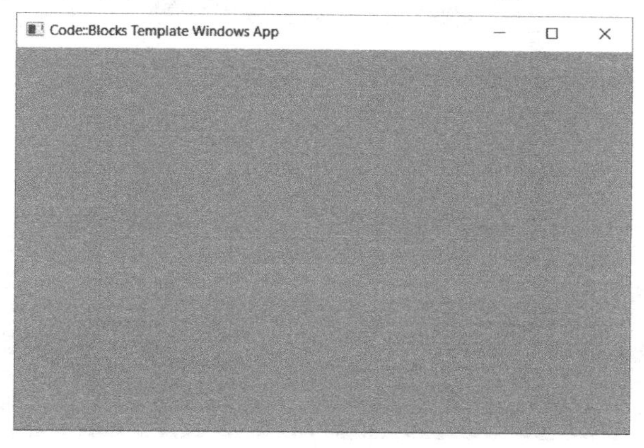

图 7-6　运行结果

7.1.2 Visual Studio 2015 第一个 Win32 GUI 程序

安装软件后，打开 Visual Studio 2015，选择“File”→“New”→“Project...”选项（如图 7-7 所示），然后在新建项目对话框中选择“Win32 Project”选项，将项目命名成“FirstTutorial”（如图 7-8 所示），在接下来的对话框中直接单击“Finish”按钮完成项目的创建。

上述操作完成后，Visual Studio 就会自动创建一些代码（如图 7-9 所示）。选择“Debug”→“Start Without Debugging”或者按快捷键 Ctrl+F5 直接对工程进行编译并运行。运行结果如图 7-10 所示。

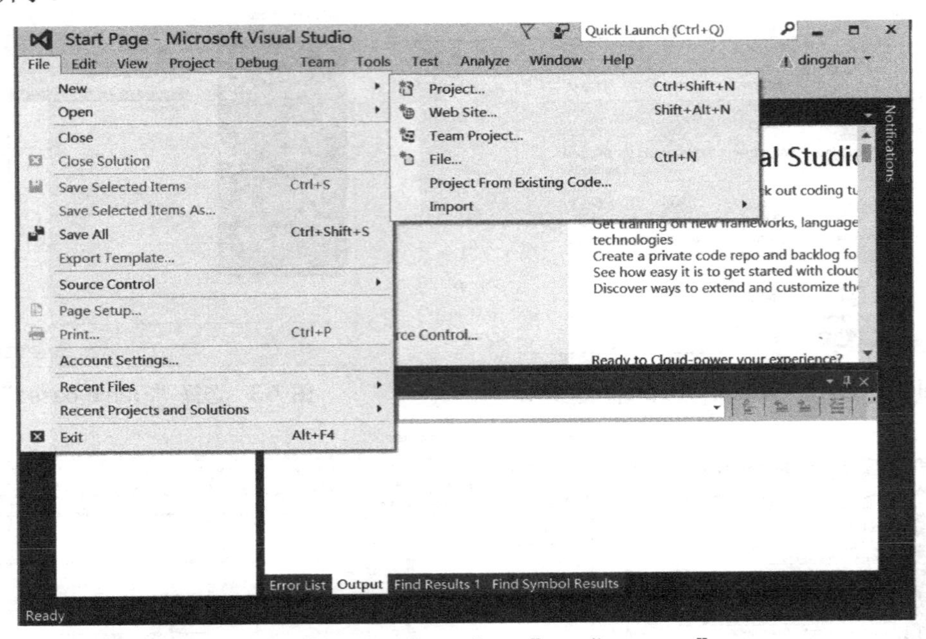

图 7-7 选择“File”→“New”→“Project...”

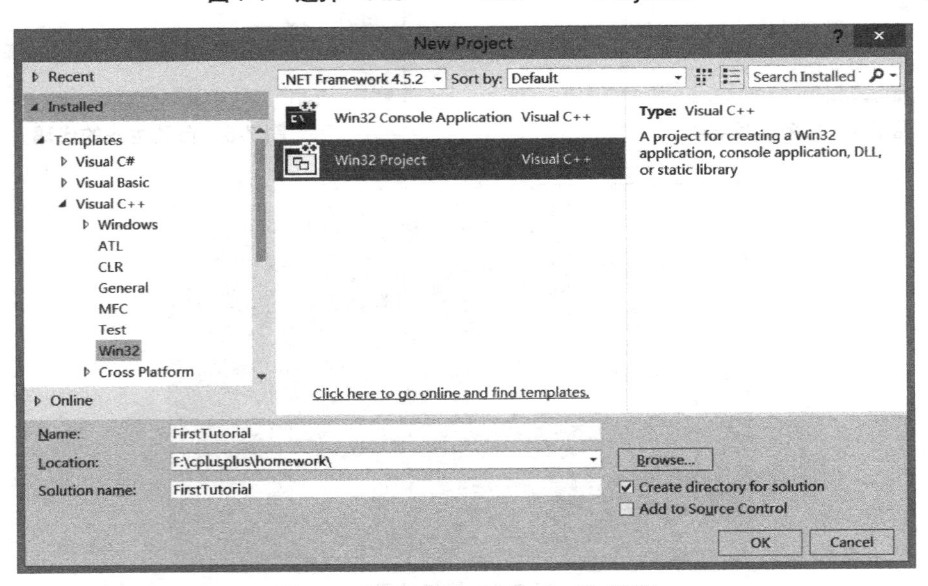

图 7-8 选择“Win32 Project”选项

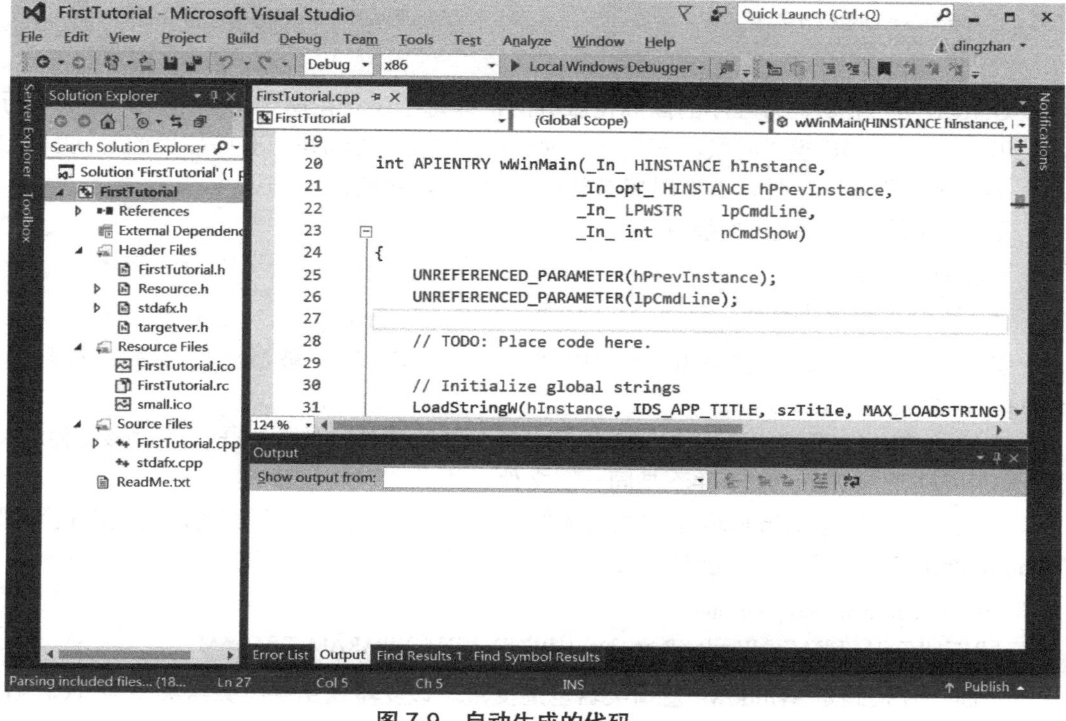

图 7-9 自动生成的代码

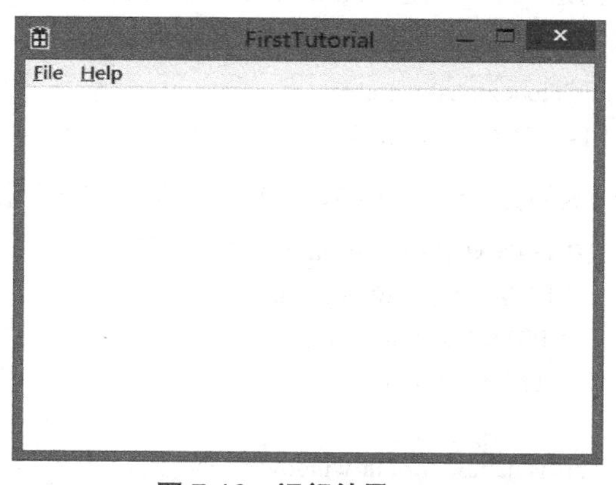

图 7-10 运行结果

7.1.3 代码分析

CodeBlocks 和 Visual Studio 2015 创建的 Win32 工程都提供初始代码，现在以 CodeBlocks 代码为例分析，Visual Studio 2015 代码功能与其类似。

以下是 5 行宏定义代码，它的功能是如果有宏 UNICODE 或_UNICODE 被声明，那么确保 UNICODE 和_UNICODE 同时被声明，代码如下：

```
#if defined(UNICODE) && !defined(_UNICODE)
    #define _UNICODE
#elif defined(_UNICODE) && !defined(UNICODE)
```

```
    #define UNICODE
#endif
```

UNICODE 主要是为程序的国际化服务的，即每个字符由两个字节组成。但由于历史原因，开启 UNICODE 模式时候，有些地方使用的宏是 UNICODE，有些使用的是_UNICODE。所以为了确保程序正确性，有些时候必须同时声明这两个宏。

下面是两行头文件包含代码：

```
#include <tchar.h>
#include <windows.h>
```

其中，tchar 是 char 和 wchar_t 的中间件，当 UNICODE 被激活时各种字符函数使用 wchar_t，否则使用 char。

在 CodeBlocks 环境中开发 Win32 程序必须包括头文件<windows.h>，不过该文件和 Visual Studio 中包括的<windows.h>并不是同一个文件。

下面一行代码是 Windows 消息处理的回调函数的声明。今后所有的消息处理代码都写在 WindowProcedure 函数内，代码如下：

```
/*  Declare Windows procedure  */
LRESULT CALLBACK WindowProcedure (HWND, UINT, WPARAM, LPARAM);
```

下面一行代码是 Windows 窗口类名全局变量。该变量使用字符数组类型为 TCHAR，而在字符串常量的前面使用宏_T。_T 是 UNICODE 和非 UNICODE 字符串常量统一表示。需要指出该字符串并不是窗口标题栏显示的文字，代码如下：


```
/*  Make the class name into a global variable  */
TCHAR szClassName[ ] = _T("CodeBlocksWindowsApp");
```

Console 程序的入口函数是 main，而 Win32 GUI 程序的入口函数是 WinMain，代码如下：

```
int WINAPI WinMain (HINSTANCE hThisInstance,
                    HINSTANCE hPrevInstance,
                    LPSTR lpszArgument,
                    int nCmdShow)
{
    HWND hwnd; /* This is handle for our window */
    MSG messages; /* Here messages to the application are saved */
    WNDCLASSEX wincl;/* Data structure for the windowclass */

    /* The Window structure */
    wincl.hInstance = hThisInstance;
    wincl.lpszClassName = szClassName;
    wincl.lpfnWndProc = WindowProcedure; /* This function is called by windows */
```

Windows 系统中，窗体实际上都是由结构为 WNDCLASSEX 的变量定义的。而其中一个很重要的工作就是设置消息处理的回调函数。所谓回调函数就是一个函数指针，wincl.lpfnWndProc = WindowProcedure 设置了该函数指针。

示例代码如下：

```
    wincl.style = CS_DBLCLKS;                      /* Catch double-clicks */
    wincl.cbSize = sizeof (WNDCLASSEX);

    /* Use default icon and mouse-pointer */
    wincl.hIcon = LoadIcon (NULL, IDI_APPLICATION);
    wincl.hIconSm = LoadIcon (NULL, IDI_APPLICATION);
    wincl.hCursor = LoadCursor (NULL, IDC_ARROW);
    wincl.lpszMenuName = NULL;                     /* No menu */
    wincl.cbClsExtra = 0;          /* No extra bytes after the window class */
    wincl.cbWndExtra = 0;                /* structure or the window instance */
    /* Use Windows's default colour as the background of the window */
    wincl.hbrBackground = (HBRUSH) COLOR_BACKGROUND;
```

上述代码设置了窗口数据结构中的各个成员，其中包括大小图标、光标、菜单、背景刷等。读者可以将上述两个 IDI_APPLICATION 都改成 IDI_ASTERISK，编译运行后，程序图标如图 7-11 所示。还可以将 IDC_ARROW 改成 IDC_CROSS，这样光标就变成了十字形。最后还可以修改程序背景色，例如，将 COLOR_BACKGROUND 改成 COLOR_WINDOW。

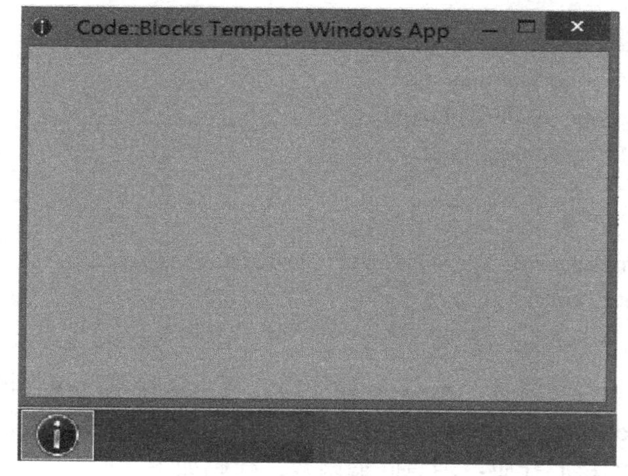

图 7-11　程序图标

当初始化窗口结构后，接着就注册窗口类并创建窗口，代码如下：

```
    /* Register the window class, and if it fails quit the program */
    if (!RegisterClassEx (&wincl))
        return 0;
    /* The class is registered, let's create the program*/
    hwnd = CreateWindowEx (        0,                  /* Extended possibilites for variation */
            szClassName,                               /* Classname */
            _T("Code::Blocks Template Windows App"),   /* Title Text */
            WS_OVERLAPPEDWINDOW,                       /* default window */
            CW_USEDEFAULT,                             /* Windows decides the position */
            CW_USEDEFAULT,                             /* where the window ends up on the screen */
            544,                                       /* The programs width */
```

```
            375,                    /* and height in pixels */
            HWND_DESKTOP,           /* The window is a child-window to desktop */
            NULL,                   /* No menu */
            hThisInstance,          /* Program Instance handler */
            NULL                    /* No Window Creation data */
            );
```

修改其中一些数据，然后观察窗口的变化。例如，修改数字544，375，这是窗口像素宽度和高度。修改"Code::Blocks Template Windows App"，这是窗口标题字符串。将WS_OVERLAPPEDWINDOW修改成WS_DLGFRAME，窗口将变成对话框模式。

创建窗口完毕后，就要显示该窗口，代码如下：

```
  /* Make the window visible on the screen */
    ShowWindow (hwnd, nCmdShow);
```

窗口中显示的代码就是Win32的消息循环，代码如下：

```
    /* Run the message loop. It will run until GetMessage() returns 0 */
    while (GetMessage (&messages, NULL, 0, 0))
    {
        /* Translate virtual-key messages into character messages */
        TranslateMessage(&messages);
        /* Send message to WindowProcedure */
        DispatchMessage(&messages);
    }

  /*The program return-value is 0 - The value that PostQuitMessage() gave */
    return messages.wParam;
  }
```


只要窗口没有收到关闭消息，该while循环将一直从操作系统接收消息，然后分配消息到回调函数WindowProcedure，并由其进行处理。

以下就是要关注的核心代码：消息回调函数，代码如下：

```
  /*  This function is called by the Windows function DispatchMessage()   */
  LRESULT CALLBACK WindowProcedure (HWND hwnd, UINT message, WPARAM wParam, LPARAM
lParam)
  {
    switch (message)                        /* handle the messages */
    {
        case WM_DESTROY:
            PostQuitMessage (0);        /* send a WM_QUIT to the message queue */
            break;
        default:                            /* for messages that we don't deal with */
            return DefWindowProc (hwnd, message, wParam, lParam);
    }
    return 0;
```

```
    }
```

初始的回调函数只处理了 WM_DESTROY 消息。WM 是 Windows Message 的简写，Win32 程序中的各类消息都是以 WM_作为前缀的整型数值。当用户单击窗口右上角的关闭按钮，这个动作会向这个消息处理函数传送一个 WM_DESTROY 消息，当接收到这个消息的时候，程序知道用户想关闭窗口。此时可以通过调用 PostQuitMessage (0)结束整个窗口的消息循环。在实际软件开发中，通常需要在窗口关闭消息分支处理中，询问用户是否需要保存修改的内容。

7.2 Win32 消息基础

Win32 GUI 程序主要通过消息机制来实现程序的交互性，因此这里需要了解各种消息的用法。

7.2.1 窗口关闭消息 WM_CLOSE

首先来看窗口关闭消息 WM_CLOSE。修改消息回调函数 WindowProcedure，代码如下：

```
switch (message)
    {
        case WM_DESTROY:
            PostQuitMessage (0);
            break;
        case WM_CLOSE:
            MessageBeep(MB_ICONASTERISK);
PlaySound(_T("C:\\WINDOWS\\Media\\Alarm01.wav"),NULL,SND_FILENAME|SND_SYNC);
            break;
        default:
            return DefWindowProc (hwnd, message, wParam, lParam);
    }
```

在 switch-case 分支中，新加入 WM_CLOSE 消息的处理功能。在第 7.1.3 节如果用户单击右上角的关闭按钮，该消息回调函数会接收 WM_DESTROY 消息，但更具体的是，它首先接收的是 WM_CLOSE 消息。在处理 WM_CLOSE 消息时，这里代码阻止了它的关闭功能，而是让它发出了 beep 声音，之后又播放一段 wav 格式的声音。

这里调用了一个库函数 PlaySound，它在头文件<mmsystem.h>中声明，库文件是 winmm.lib。遗憾的是 gcc/g++不支持 #pragma comment（lib, “WINMM.LIB”） 方式导入静态库，所以需手动向 CodeBlocks 中导入该库，如图 7-12 所示，具体导入方法如下。

（1）右击“FirstTutorial”工程，在弹出的菜单中选择“Properties”选项。

（2）再单击“Project’s build options...”按钮。

（3）在接下来的对话框中选择“Linker settings”选项。

（4）单击“Add”按钮，在弹出的对话框中输入“winmm”，单击“OK”按钮。

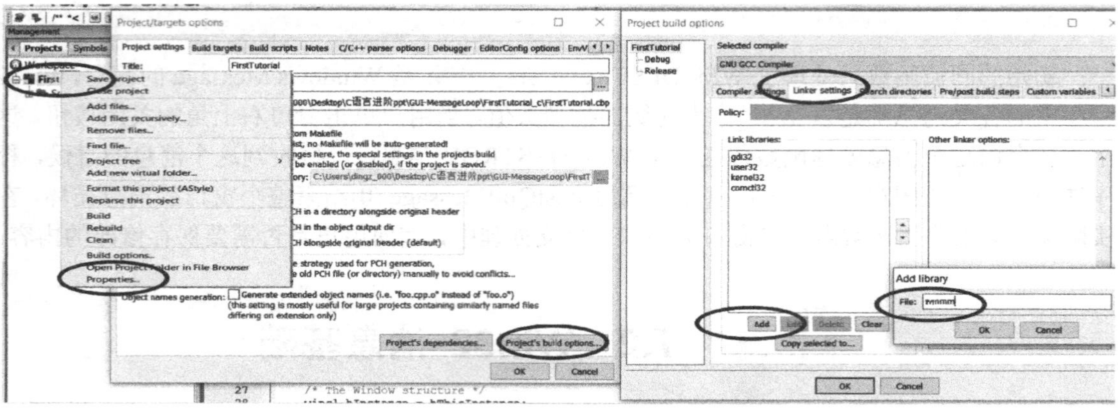

图 7-12　手动向 CodeBlocks 中导入 winmm.lib

试试看，单击关闭按钮时是否能发出声音？

7.2.2　窗口大小调整消息 WM_SIZE

继续修改 WindowProcedure 函数，添加对 WM_SIZE 消息的响应，代码如下：


```
RECT rect;
...
switch (message)
{
...
case WM_SIZE:
    ::GetWindowRect(hwnd, &rect); //得到窗口尺寸
    if(rect.right-rect.left>800)
PlaySound(_T("C:\\WINDOWS\\Media\\Alarm01.wav"),NULL,SND_FILENAME|SND_ASYNC);
    else if(rect.right-rect.left>400)
PlaySound(_T("C:\\WINDOWS\\Media\\Alarm02.wav"),NULL,SND_FILENAME|SND_ASYNC);
    else
PlaySound(_T("C:\\WINDOWS\\Media\\Alarm03.wav"),NULL,SND_FILENAME|SND_ASYNC);
    break;
```

只要窗口的大小发生了变化，回调函数一定会收到若干 WM_SIZE 消息。在上述程序中，用户拖曳窗口边缘从而调整窗口大小时，程序会播放不同的声音。窗口宽度大于 800 像素播放 Alarm01，小于 800 并大于 400 播放 Alarm02，其他时候播放 Alarm03。

上述 GetWindowRect 函数用来获得窗口的大小，第一个参数是窗口句柄，也就是当前这个窗口的身份证，不同的窗口有不同的句柄。第二个参数是 RECT 结构类型的指针，用来获得窗口矩形大小。GetWindowRect 函数声明如下：

```
BOOL GetWindowRect( HWND hWnd, LPRECT lpRect );
```

7.2.3 窗口创建消息 WM_CREATE

每个窗口都有一个生命周期，而窗口的创建会产生一个 WM_CREATE 消息。这里希望在接收 WM_CREATE 消息时创建窗口的菜单。

和 Visual Studio 不同的是，CodeBlocks 中无法通过可视化的资源编辑工具创建或修改菜单栏。下面通过手动输入代码的方式定义 CodeBlocks 窗口下的菜单栏，代码如下：

```
#define IDM_FILE_NEW    1
#define IDM_FILE_OPEN   2
#define IDM_FILE_QUIT   3

#define IDM_HELP_HELPDOC    4
#define IDM_HELP_ABOUTUS    5

//创建主窗口菜单：菜单栏→菜单→菜单项
void MakeMainMenu(HWND hWnd)
{
    HMENU hMenuBar;         // 菜单栏
    HMENU hMenuFile;        // File 菜单
    HMENU hMenuHelp;        // Help 菜单
    //创建菜单栏
    hMenuBar = CreateMenu();

    //创建菜单 File
    hMenuFile = CreateMenu();
    //在菜单 File 下，添加菜单项 New
    AppendMenuW(hMenuFile, MF_STRING, IDM_FILE_NEW, L"&New");
    //在菜单 File 下，添加菜单项 Open
    AppendMenuW(hMenuFile, MF_STRING, IDM_FILE_OPEN, L"&Open");
    //在菜单 File 下，添加一条分割线
    AppendMenuW(hMenuFile, MF_SEPARATOR, 0, NULL);
    //在菜单 File 下，添加菜单项 Quit
    AppendMenuW(hMenuFile, MF_STRING, IDM_FILE_QUIT, L"&Quit");

    //将菜单 File 添加到主菜单栏
    AppendMenuW(hMenuBar, MF_POPUP, (UINT_PTR)hMenuFile, L"&File");

     //创建菜单 Help
    hMenuHelp = CreateMenu();
    //在菜单 Help 下，添加菜单项 Help Document
    AppendMenuW(hMenuHelp, MF_STRING, IDM_HELP_HELPDOC, L"&Help Document");
    //在菜单 Help 下，添加菜单项 About us
    AppendMenuW(hMenuHelp, MF_STRING, IDM_HELP_ABOUTUS, L"&About us");
```

```
        //将菜单 Help 添加到主菜单栏
        AppendMenuW(hMenuBar, MF_POPUP, (UINT_PTR)hMenuHelp, L"&Help");

        //设置当前窗口的菜单栏
        SetMenu(hWnd, hMenuBar);
        return;
}
```

读者不需要了解窗口菜单的工作机理，但要知道一个窗口可以关联一个菜单栏，每个菜单栏下可以有多个菜单，而每个菜单下又可以定义多个菜单项。所以，手动编写菜单时，必须定义三项内容：菜单栏→菜单→菜单项，最后不要忘记将菜单栏关联到窗口。读者可以将上述代码定义在 WindowProcedure 之前。下面就是要找一个合适的地方调用上述 MakeMainMenu 函数。这里，在 WindowProcedure 函数中响应 WM_CREATE 消息即可，代码如下：

```
        case WM_CREATE:
            //创建菜单
            MakeMainMenu(hwnd);
            break;
```

7.2.4 菜单或其他按钮消息 WM_COMMAND

使用上述代码可以让菜单成功出现在窗口中。下面需要响应每个菜单的消息，代码如下：

```
    case WM_COMMAND:
        wmId= LOWORD(wParam);                //被单击菜单的 ID
        wmEvent = HIWORD(wParam);            //被单击对象的事件标志
        // Parse the menu selections:
        switch (wmId)
        {
        case IDM_FILE_NEW:
     PlaySound(_T("C:\\WINDOWS\\Media\\Alarm01.wav"), NULL,SND_FILEN AME|SND_ASYNC);
            break;
        case IDM_FILE_OPEN:
     PlaySound(_T("C:\\WINDOWS\\Media\\Alarm02.wav"), NULL,SND_FILEN AME|SND_ASYNC);
            break;
        case IDM_FILE_QUIT:
            DestroyWindow(hwnd);
            break;
        case IDM_HELP_HELPDOC:
    PlaySound(_T("C:\\WINDOWS\\Media\\Alarm03.wav"), NULL,SND_FILEN
AME|SND_ASYNC);
            break;
        case IDM_HELP_ABOUTUS:
            MessageBeep(MB_ICONASTERISK);
```

```
            break;
        default:
            break;
        }
        break;
```

首先，任何一个菜单项被单击后发送的消息都是 WM_COMMAND。程序需要通过 wParam 的低字节值获得具体菜单的 ID 号，这里使用宏 LOWORD 获得 wParam 的低字节值。

每个菜单项被单击后，都调用 PlaySound 播放了一段音乐。如果想让某个音频内容成为背景音乐，那么需要通过如下方式调用 PlaySound，代码如下：

```
PlaySound(_T("C:\\WINDOWS\\Media\\Alarm01.wav"),NULL,SND_FILENAME|SND_ASYNC|SND_LOOP);
```

SND_ASYNC 表示采用异步方式播放声音，SND_LOOP 表示循环播放。那么，如何控制背景音乐的开与关呢？下面是一段不难理解的伪代码：

```
    static int s_bPlaying = 0;
    if(s_bPlaying )
        PlaySound(…);
    else
        PlaySound(NULL, NULL, 0);
    s_bPlaying = s_bPlaying?0:1;
```


7.2.5 鼠标消息

鼠标消息包括鼠标左键消息：

```
WM_LBUTTONDOWN        鼠标左键按下
WM_LBUTTONUP          鼠标左键弹起
```

鼠标右键消息：

```
WM_RBUTTONDOWN        鼠标右键按下
WM_RBUTTONUP          鼠标右键弹起
```

鼠标移动消息：

```
WM_MOUSEMOVE          鼠标移动
```

鼠标双击消息：

```
WM_LBUTTONDBLCLK      鼠标左键双击
WM_RBUTTONDBLCLK      鼠标右键双击
```

鼠标滚轮消息

```
WM_MOUSEWHEEL         鼠标中键滚动
```

在鼠标消息中，一个很重要的问题是如何获得当前鼠标位置？这里使用两个宏得到鼠标位置，代码如下：

```
xPos = GET_X_LPARAM(lParam);
yPos = GET_Y_LPARAM(lParam);
```

为了使用这两个宏，需要引入头文件<Windowsx.h> 。

下面继续往 WindowProcedure 回调函数中添加鼠标响应的代码，代码如下：

```
        static POINT s_PointDown, s_PointUp;
        int zDelta;
    static bool bHasLine=false;
        ...
        case WM_LBUTTONDOWN:
            s_PointDown.x = GET_X_LPARAM(lParam);
            s_PointDown.y = GET_Y_LPARAM(lParam);
            break;
        case WM_LBUTTONUP:
            s_PointUp.x = GET_X_LPARAM(lParam);
            s_PointUp.y = GET_Y_LPARAM(lParam);
            bHasLine = true;
            ::InvalidateRect(hwnd, NULL, 1);
            break;
        case WM_MOUSEWHEEL:
            zDelta = GET_WHEEL_DELTA_WPARAM(wParam);
            if(bHasLine)
            {
                if(zDelta>0)
                {
                    s_PointDown.y+=10;
                    s_PointUp.y+=10;
                }
                else
                {
                    s_PointDown.y-=10;
                    s_PointUp.y-=10;
                }
                ::InvalidateRect(hwnd, NULL, 1);
            }
            break;
```

上面代码定义了两个静态 POINT 结构成员变量，一个保存鼠标左键按下的位置，另一个保存鼠标左键弹起的位置。而鼠标滚轮消息对这两个点进行了上下平移。这里还使用了 InvalidateRect，这是用来主动刷新屏幕的系统函数。加入上述代码后，屏幕上没有显示任何东西。

7.2.6 绘制消息 WM_PAINT

在 Windows 系统中，窗口绘制消息是 WM_PAINT。所有需要绘制的图形、图像内容最好都放在 WM_PAINT 的消息响应代码中。这里只需要绘制上一节的那条线段，代码如下：

```
HDC hdc;
PAINTSTRUCT ps;
....
```

```
case WM_PAINT:
    hdc = BeginPaint(hwnd, &ps);
    if(bHasLine)
    {
        MoveToEx(hdc, s_PointDown.x, s_PointDown.y, NULL);
        LineTo(hdc, s_PointUp.x, s_PointUp.y);
    }
    EndPaint(hwnd, &ps);
    break;
```

任何需要绘制的内容请放在 BeginPaint 和 EndPaint 代码之间。hdc 是绘制设备的句柄，ps 是绘图结构。真正的绘制代码是 MoveToEx 和 LineTo。其中 MoveToE 表示将起始绘制点移动到上一次鼠标按下的位置，而 LineTo 表示将绘制线段的终点设置在鼠标左键弹起的位置。把上述代码加入程序后，就能用鼠标绘制线段。而且还能通过滚动鼠标中键上下平移线段。

有些时候程序代码需要主动激活 WM_PAINT 消息，此时可以使用::InvalidateRect(hWnd, NULL, 1);表示刷新整个窗口。

7.2.7 键盘消息 WM_KEYDOWN 和 WM_KEYUP

键盘消息 WM_KEYDOWN 表示键盘中某键被按下， WM_KEYUP 表示某个键被弹起。如何得到哪个键被按下了？答案是利用 wParam 参数，代码如下：

```
case WM_KEYDOWN:
    switch (wParam)
    {
    case VK_LEFT:
        if(bHasLine)
        {
            s_PointDown.x-=10;
            s_PointUp.x-=10;
            ::InvalidateRect(hwnd, NULL, 1);
        }
        break;
    case VK_RIGHT:
        if(bHasLine)
        {
            s_PointDown.x+=10;
            s_PointUp.x+=10;
            ::InvalidateRect(hwnd, NULL, 1);
        }
        break;
    case VK_UP:
        break;
    case VK_DOWN:
```

```
            break;
        case 'A':
        case 'a':
            break;
        }
        break;
```

上述代码只处理了方向键中的左键和右键，它可以对当前绘制的线段左右平移。方向键属于虚拟按键，所以是 VK_开头的宏(virtual key)。而对于普通的按键，就直接使用该键的 ASCII 码。请读者参考上面的 'A' 和 'a' 写法。

有时需要获取 Shift 键是否被按下，则使用下面的代码：

```
iState = HIWORD(GetKeyState(VK_SHIFT));
```

或者 Control 键是否被按下，代码如下：

```
iState = HIWORD(GetKeyState(VK_CONTROL));
```

其中，GetKeyState 用来获得任何按键的状态，如果想得到一个按键是否被按下，需要使用宏 HIWORD 获得 GetKeyState 的高字节部分，如果该高字节值为 1，则表示被按下。

7.3　综合编程实例：简单多边形的创建、绘制、平移与旋转

第 7.2 节已经介绍 Win32 消息中若干重要的内容，下面使用一个实例作为练习，手动编写的代码量约为 300 行。实例任务如下：

（1）创建多边形：用户单击视图区后就开始创建多边形，鼠标右键弹起后将该多边形封闭，同时表示该多边形创建结束。

（2）绘制多边形：在 WM_PAINT 消息中绘制创建所有多边形。

（3）清除多边形：利用菜单项 Clear，清除当前视图中所有多边形。

（4）平移多边形：利用 Ctrl+鼠标左键，平移视图中所有多边形。

（5）旋转多边形：利用键盘 r 或 R 键，旋转视图中所有多边形。

（6）Undo：单击菜单中的 Undo 按钮或者使用快捷键 Ctrl+Z，撤销上一次的操作。

（7）保存多边形：利用菜单项 Save，将当前视图中所有的多边形保存到 a.prj 中。

（8）载入多边形文件 a.prj：利用菜单项 Open，将 a.prj 文件载入后，能够正常显示和编辑多边形。

1. 创建点、多边形和多边形管理器类

当读者从 Console 程序迁移到 GUI 程序后，需要将函数和类存放在一个源文件的习惯丢弃，多文件结构是实际工程中最常用的写法。这里将数据部分和图形界面部分分离，数据部分放入新建的 polygon.h 和 polygon.cpp 文件。

因为涉及图形，所以需要定义基本的二维点，虽然 GDI 中提供了 POINT 结构，但是它的两个坐标分量都是整型。整型坐标点在旋转后会失真，所以这里定义一个二维点类 Point2f，

它的两个坐标分量都采用浮点类型。同时，利用 STL 的 std::vector 将点打包成 MyPolygon 类型，该类包含了多边形的各项操作。因为屏幕中有多个多边形，所以还需要一个多边形管理器。另外，考虑程序需要设置撤销操作，所以这里还要定义撤销数据类型。基于上述描述，Polygon.h 中四个类声明如下。其中 Point2f 类、定义二维点类、点坐标是浮点类型，这是考虑多边形旋转后不会变形，代码如下：

```
class Point2f
{
public:
    double x;
    double y;
    Point2f(double xx=0, double yy=0):x(xx),y(yy)
    { }
    //平移
    void Translate(double dx, double dy)
    {
        x+=dx;
        y+=dy;
    }
    //旋转公式，高中数学内容，需要注意旋转中心位于窗口中心，而不是原点(0,0)
    void Rotate(double cx, double cy, double angle)
    {
        double xx, yy;
        xx = (x-cx)*cos(angle)-(y-cy)*sin(angle)+cx;
        yy = (x-cx)*sin(angle)+(y-cy)*cos(angle)+cy;
        x = xx;
        y = yy;
    }
};
```

定义多边形类 MyPolygon，代码如下：

```
class MyPolygon
{
    public:
        MyPolygon();
        virtual ~MyPolygon();
        //添加一个新点
        void AddPoint(const Point2f& pnt) {m_points.push_back(pnt);}
        //平移
        void Translate(double dx, double dy);
        //旋转
        void Rotate(const Point2f& center, double degree);
        //是否为空
        bool IsEmpty() { return m_points.empty();}
```

```
        //保存
        bool Save(std::ofstream& file) const;
        //载入
        bool Load(std::ifstream& file);
        //删除最后一点
        bool RemoveLastPoint();
        //绘制
        void Draw(HDC hdc);
        //当前多边形点数
        size_t GetNumPoints() {return m_points.size();}
        //清除所有点
        void Clear() {m_points.clear();}
        //封闭多边形
        bool MakeClose();
    private:
        std::vector<Point2f> m_points;
};
```

定义撤销数据类 UndoData，代码如下：


```
class UndoData
{
public:
    //包括了三种类型：撤销创建的多边形，撤销平移，撤销旋转
    enum UndoType{UT_ADD_POLYGON, UT_TRANSLATE, UT_ROTATE};
public:
    UndoType m_UndoType;        //Undo 类型
    double m_Translate_dx;      //UT_TRANSLATE only
    double m_Translate_dy;      //UT_TRANSLATE only
    double m_Rotate_Degree;     //UT_ROTATE only
    Point2f m_Rotate_Center;    //UT_ROTATE only
};
```

定义多边形管理器类 PolygonManager，代码如下：

```
class PolygonManager
{
public:
    PolygonManager();
    //添加一个点
    void AddPoint(int x, int y);
    //设置当前编辑的多边形最后一个点
    bool SetMovingPoint(int x, int y);
    //添加一个多边形
    void AddPolygon();
    //撤销
```

```
    bool Undo();
    //清除所有多边形，同时清除 Undo 堆栈
    void Clear();
    //平移，最后一个参数表示该平移是否进入 Undo 堆栈
    void Translate(double dx, double dy, bool bToUndoStack);
    //旋转，最后一个参数表示该旋转是否进入 Undo 堆栈
    void Rotate(const Point2f& center, double degree, bool bToUndoStack);
    //保存
    bool Save(const std::string& fileName) const;
    //载入
    bool Load(const std::string& fileName);
    //绘制
    void Draw(HDC hdc);
private:
    std::vector<MyPolygon> m_Polygons;       //已经完成的多边形
    MyPolygon m_CurrentPolygon;              //当前正在创建的多边形
    Point2f m_MovingPoint;                   //当前正在创建的多边形的移动点
    std::stack<UndoData> m_UndoStack;        //Undo 堆栈
    double m_Translate_Sum_X;       //平移的总的 X 量，for Translate Undo only
    double m_Translate_Sum_Y;       //平移的总的 Y 量，for Translate Undo only
    double m_Rotate_Sum_Degree;     //旋转的总的角度量，for Rotate Undo only
};
```


Polygon.cpp 中重要函数定义如下：

```
//平移
void MyPolygon::Translate(double dx, double dy)
{
    if(m_points.empty())
        return;
    size_t i;
    for(i=0; i<m_points.size(); ++i)
        m_points[i].Translate(dx, dy);
}
//旋转
void MyPolygon::Rotate(const Point2f& center, double degree)
{
    if(m_points.empty())
        return;
    size_t i;
    //角度转弧度
    double angle = DEGREE_TO_ANGLE(degree);
    for(i=0; i<m_points.size(); ++i)
        m_points[i].Rotate(center.x, center.y, angle);
}
//保存，采用二进制保存方式
```

```
bool MyPolygon::Save(std::ofstream& file) const
{
    size_t i, n;
    n = m_points.size();
    //先保存点数
    file.write((char*)(&n), sizeof(n));
    //再保存每个点的 X 和 Y 坐标
    for(i=0; i<m_points.size(); ++i)
    {
        file.write((char*)(&m_points[i].x),sizeof(m_points[i].x));
        file.write((char*)(&m_points[i].y),sizeof(m_points[i].y));
    }
    return true;
}
//载入，采用二进制载入方式
bool MyPolygon::Load(std::ifstream& file)
{
    size_t i, n;
    Point2f pnt;
    m_points.clear();
    //先读取点数
    file.read((char*)(&n), sizeof(n));
    if(file.fail())
        return false;
    //如果数据量异常则退出
    if(n<0)
        return false;
    //再读取所有点的 X 和 Y 坐标，任何一次读取行为都应该做失败检测
    for(i=0; i<n; ++i)
    {
        file.read((char*)(&pnt.x),sizeof(pnt.x));
        file.read((char*)(&pnt.y),sizeof(pnt.y));
        if(file.fail())
            return false;
        m_points.push_back(pnt);
    }
    return true;
}

//删除最后一个点
bool MyPolygon::RemoveLastPoint()
{
    if(m_points.empty())
```

```
            return false;
        m_points.pop_back();
        return true;
}
//封闭多边形
bool MyPolygon::MakeClose()
{
        if(m_points.size()<2)
            return false;
        if(m_points.size()= =2)
            return true;
        m_points.push_back(m_points.front());
        return true;
}
//绘制
void MyPolygon::Draw(HDC hdc)
{
        if(m_points.size()<2)
            return;
        size_t i;
        ::MoveToEx(hdc, (int)m_points[0].x, (int)m_points[0].y , NULL);
        for(i=1; i<m_points.size(); ++i)
            ::LineTo(hdc, (int)m_points[i].x, (int)m_points[i].y);
}

//构造函数，将 Undo 需要的数据先初始化
PolygonManager::PolygonManager()
{
        m_Translate_Sum_X = 0;
        m_Translate_Sum_Y = 0;
        m_Rotate_Sum_Degree = 0;
}

//往当前编辑的多边形中加入一个点
void PolygonManager::AddPoint(int x, int y)
{
        m_CurrentPolygon.AddPoint(Point2f(x, y));
}

//当前编辑的多边形中设置移动点
bool PolygonManager::SetMovingPoint(int x, int y)
{
        m_MovingPoint.x = x;
        m_MovingPoint.y = y;
```

```
        return !m_CurrentPolygon.IsEmpty();
}

//添加多边形
void PolygonManager::AddPolygon()
{
        //先把当前编辑的多边形封闭
        if(!m_CurrentPolygon.MakeClose())
        {
                m_CurrentPolygon.Clear();
                return;
        }
        m_Polygons.push_back(m_CurrentPolygon);
        //添加完成后，记住要清除当前编辑的多边形
        m_CurrentPolygon.Clear();

        //添加本次动作到 Undo 堆栈
        UndoData ud;
        ud.m_UndoType = UndoData::UT_ADD_POLYGON;
        m_UndoStack.push(ud);
}

//撤销上一次动作
bool PolygonManager::Undo()
{
        //如果有正在编辑的多边形，那么先撤销该多边形的最后一个点
        if(!m_CurrentPolygon.IsEmpty())
        {
                m_CurrentPolygon.RemoveLastPoint();
                return true;
        }
        //如果堆栈为空，没有可撤销的内容
        if(m_UndoStack.empty())
                return false;

        size_t i;
        //得到需要撤销的数据
        UndoData ud = m_UndoStack.top();
        //弹出堆栈
        m_UndoStack.pop();
        switch(ud.m_UndoType)
        {
        case UndoData::UT_ADD_POLYGON:
```

```
            //撤销添加的多边形
            if(m_Polygons.empty())
                return false;
            m_Polygons.pop_back();
            break;
        case UndoData::UT_TRANSLATE:
            //撤销平移动作
            {
                for(i=0; i<m_Polygons.size(); ++i)
                    m_Polygons[i].Translate(-ud.m_Translate_dx, -ud.m_Translate_dy);
            }
            break;
        case UndoData::UT_ROTATE:
            //撤销旋转动作
            {
                for(i=0; i<m_Polygons.size(); ++i)
                    m_Polygons[i].Rotate(ud.m_Rotate_Center, -ud.m_Rotate_Degree);
            }
            break;
    }
    return true;
}

//清除所有多边形及 Undo 堆栈
void PolygonManager::Clear()
{
    m_Polygons.clear();
    while(!m_UndoStack.empty())
        m_UndoStack.pop();
    m_Translate_Sum_X = 0;
    m_Translate_Sum_Y = 0;
    m_Rotate_Sum_Degree = 0;
}

//平移
void PolygonManager::Translate(double dx, double dy, bool bToUndoStack)
{
    size_t i;
    for(i=0; i<m_Polygons.size(); ++i)
        m_Polygons[i].Translate(dx, dy);
    //一定要累加平移量
    m_Translate_Sum_X+=dx;
    m_Translate_Sum_Y+=dy;
```

```
    //如果需要加入堆栈
    if(bToUndoStack)
    {
        UndoData ud;
        ud.m_UndoType = UndoData::UT_TRANSLATE;
        ud.m_Translate_dx = m_Translate_Sum_X;
        ud.m_Translate_dy = m_Translate_Sum_Y;
        m_UndoStack.push(ud);
        //把累加的平移量清空为 0
        m_Translate_Sum_X = 0;
        m_Translate_Sum_Y = 0;
    }
}

//旋转
void PolygonManager::Rotate(const Point2f& center, double degree, bool bToUndoStack)
{
    size_t i;
    for(i=0; i<m_Polygons.size(); ++i)
        m_Polygons[i].Rotate(center, degree);
    //一定要累加旋转的角度量
    m_Rotate_Sum_Degree+=degree;
    //如果需要加入堆栈
    if(bToUndoStack)
    {
        UndoData ud;
        ud.m_UndoType = UndoData::UT_ROTATE;
        ud.m_Rotate_Degree = m_Rotate_Sum_Degree;
        ud.m_Rotate_Center = center;
        m_UndoStack.push(ud);
        //把累加的旋转量清空为 0
        m_Rotate_Sum_Degree = 0;
    }
}

//保存文件
bool PolygonManager::Save(const std::string& fileName) const
{
    std::ofstream file(fileName.c_str(), std::ios::binary);
    size_t i, n;
    if(file.fail())
        return false;
    n = m_Polygons.size();
```

```
    //先保存多边形个数
    file.write((char*)(&n), sizeof(n));
    //保存每个多边形数据
    for(i=0; i<n; ++i)
        m_Polygons[i].Save(file);
    return true;
}

//读入文件
bool PolygonManager::Load(const std::string& fileName)
{
    Clear();

    std::ifstream file(fileName.c_str(), std::ios::binary);
    MyPolygon polygon;
    size_t i, n;
    if(file.fail())
        return false;
    //先读入保存的多边形个数
    file.read((char*)(&n), sizeof(n));
    //如果数据量异常则退出
    if(n<0)
    {
        file.close();
        return false;
    }
    for(i=0; i<n; ++i)
    {
        //如果任何多边形读入有错，则后续就不要读取
        if(!polygon.Load(file))
        {
            file.close();
            return false;
        }
        m_Polygons.push_back(polygon);
    }
    file.close();
    return true;
}

//绘制已有的多边形和当前编辑的多边形
void PolygonManager::Draw(HDC hdc)
{
    size_t i;
```

```
        if(m_CurrentPolygon.GetNumPoints())
        {
            //绘制当前编辑的多边形，先添加一个移动点，绘制完再删除
            m_CurrentPolygon.AddPoint(m_MovingPoint);
            m_CurrentPolygon.Draw(hdc);
            m_CurrentPolygon.RemoveLastPoint();
        }
        for(i=0; i<m_Polygons.size(); ++i)
            m_Polygons[i].Draw(hdc);
    }
```

2. 多边形管理器类的使用

多边形管理器类是数据的核心，这样需要在 main.cpp 文件中定义一个全局的多边形管理器对象，然后各种操作应用在该对象即可。这里节选部分代码供读者参考，代码如下：

```
    LRESULT CALLBACK WindowProcedure (HWND hwnd, UINT message, WPARAM wParam, LPARAM
lParam)
    {
        //上一次采样的鼠标位置，用局部静态变量
        static POINT s_LastPos;
        //当前鼠标位置
        POINT curPos;
        //鼠标左键是否被按下，Control 键是否被按下
        static bool bLButtonDown=false, bCtrlDown=false;
        //旋转角度步长定义为 5 度
        static const double degreeStep = 5.0;
        //设备上下文句柄
        HDC hdc;
        //绘制结构体
        PAINTSTRUCT ps;
        //定义一个二维点
        Point2f pnt2f;
        //矩形
        RECT rect;

        switch (message)                    /* handle the messages */
        {
        case WM_CREATE:
            //创建主菜单
            CreateMainMenu(hwnd);
            break;
        case WM_COMMAND:
        {
            //各种菜单项消息
```

```
            switch(wParam)
            {
            case IDM_FILE_OPEN:
                //读入文件 a.prj
                s_polygonManager.Load("a.prj");
                InvalidateRect(hwnd, NULL, 1);
                break;
            case IDM_FILE_SAVE:
                //写入文件 a.prj
                s_polygonManager.Save("a.prj");
                break;
            case IDM_FILE_QUIT:
                //程序退出
                DestroyWindow(hwnd);
                break;
            case IDM_EDIT_UNDO:
                //撤销上一次动作
                s_polygonManager.Undo();
                InvalidateRect(hwnd, NULL, 1);
                break;
            case IDM_EDIT_CLEAR:
                //清除所有多边形
                s_polygonManager.Clear();
                InvalidateRect(hwnd, NULL, 1);
                break;
            }
        }
        break;
        case WM_DESTROY:
            PostQuitMessage (0);            /* send a WM_QUIT to the message queue */
            break;
        case WM_LBUTTONDOWN:        //鼠标左键被按下
            bLButtonDown = true;
            bCtrlDown = HIWORD(GetKeyState(VK_CONTROL));
            s_LastPos.x = GET_X_LPARAM(lParam);
            s_LastPos.y = GET_Y_LPARAM(lParam);
            //if(!bCtrlDown)
            //      s_polygonManager.AddPoint(s_LastPos.x, s_LastPos.y);
            InvalidateRect(hwnd, NULL, 1);
            break;
        case WM_MOUSEMOVE:          //鼠标移动时
            curPos.x = GET_X_LPARAM(lParam);
            curPos.y = GET_Y_LPARAM(lParam);
```

```
            //鼠标移动时，如果左键和 Control 键同时被按下，则是平移手势
            if(bLButtonDown && bCtrlDown)
            {
                s_polygonManager.Translate(curPos.x-s_LastPos.x, curPos.y-s_LastPos.y, false);
                InvalidateRect(hwnd, NULL, 1);
            }
            else
            {
                //设置当前创建的多边形的移动点
                if(s_polygonManager.SetMovingPoint(curPos.x, curPos.y))
                    InvalidateRect(hwnd, NULL, 1);
            }
            //刷新上一次坐标点
            s_LastPos = curPos;
            break;
        case WM_LBUTTONUP:
            curPos.x = GET_X_LPARAM(lParam);
            curPos.y = GET_Y_LPARAM(lParam);
            //鼠标弹起时，如果左键和 Control 键同时被按下，结束平移手势，此时需要将本次平移操作
发送到 Undo 栈
            if(bLButtonDown && bCtrlDown)
            {
                s_polygonManager.Translate(curPos.x-s_LastPos.x, curPos.y-s_LastPos.y, true);
            }
            else
            {
                //为当前创建的多边形添加一个新的点
                s_polygonManager.AddPoint(curPos.x, curPos.y);
            }
            //清除鼠标左键按下状态和 Control 键按下状态
            bLButtonDown = false;
            bCtrlDown = false;
            InvalidateRect(hwnd, NULL, 1);
            break;
        case WM_RBUTTONUP:
            //鼠标右键弹起表示结束多边形创建的手势
            s_polygonManager.AddPolygon();
            InvalidateRect(hwnd, NULL, 1);
            break;
        case WM_SIZE:
            //文档窗口的大小发生变化时，需要重新绘制整个窗口
            InvalidateRect(hwnd, NULL, 1);
            break;
```

```
    case WM_PAINT:
        hdc = BeginPaint(hwnd, &ps);
        //在窗口中心位置绘制一个参考小矩形，旋转中心位于该矩形
        GetClientRect(hwnd, &rect);
        pnt2f.x = (rect.left+rect.right)*0.5;
        pnt2f.y = (rect.top+rect.bottom)*0.5;
        Rectangle(hdc, pnt2f.x-1, pnt2f.y-1, pnt2f.x+1, pnt2f.y+1);
        //绘制多边形编辑管理器
        s_polygonManager.Draw(hdc);
        EndPaint(hwnd, &ps);
        break;

    case WM_KEYDOWN:
        switch (wParam)
        {
        case VK_LEFT:
        break;
        case VK_RIGHT:
        break;
        case VK_UP:
            break;
        case VK_DOWN:
            break;
        case 'R':
        case 'r':
            //旋转中心位于窗口中心
            GetClientRect(hwnd, &rect);
            pnt2f.x = (rect.left+rect.right)*0.5;
            pnt2f.y = (rect.top+rect.bottom)*0.5;
            s_polygonManager.Rotate(pnt2f, degreeStep, false);
            InvalidateRect(hwnd, NULL, 1);
            break;
        }
        break;
    case WM_KEYUP:
        switch (wParam)
        {
        case VK_LEFT:
            break;
        case VK_RIGHT:
            break;
        case VK_UP:
            break;
        case VK_DOWN:
```

```
                break;
            case 'R':
            case 'r':
                //旋转结束后，需要将本次动作发送到 Undo 栈中
                GetClientRect(hwnd, &rect);
                pnt2f.x = (rect.left+rect.right)*0.5;
                pnt2f.y = (rect.top+rect.bottom)*0.5;
                s_polygonManager.Rotate(pnt2f, degreeStep, true);
                InvalidateRect(hwnd, NULL, 1);
                break;
            case 'Z':
            case 'z':
                //Ctrl+Z 是 Undo 的快捷键
                bCtrlDown = HIWORD(GetKeyState(VK_CONTROL));
                if(bCtrlDown)
                {
                    s_polygonManager.Undo();
                    InvalidateRect(hwnd, NULL, 1);
                }
                break;
            }
            break;
        default:                    /* for messages that we don't deal with */
            return DefWindowProc (hwnd, message, wParam, lParam);
        }

        return 0;
}
```

读者可以从网站下载该代码，通过阅读具体代码、程序运行及调试帮助理解。

7.4 Win32 图形设备接口 GDI

Win32 GUI 程序中一般采用图形设备接口 GDI(Graphics Device Interface)进行绘制。程序设计中经常可以看到接口(Interface)这个名词，实际就是表示一系列系统库函数。Win32 GDI 库函数包含以下内容。

- 线段和曲线绘制函数。
- 笔、画刷、填充绘制函数。
- 字体和文本绘制函数。
- 光栅操作函数。
- 图像函数。

本书不是 GDI 的参考手册，所以只列出和综合课程相关的一些内容作为介绍，具体函数读者可以搜索 MSDN 进行在线查阅。

7.4.1 线段和曲线绘制

线段绘制时，首先要让设备句柄(Handle of Device Context，HDC)明确落笔点，所以需要先调用 MoveToEx 函数。MoveToEx 最后一个参数返回的是上一次笔的位置，代码如下：

```
BOOL MoveToEx(
    _In_    HDC         hdc,
    _In_    int         X,
    _In_    int         Y,
    _Out_ LPPOINT lpPoint
);
```

有了落笔点，接着就开始画线段，这就是 LineTo 函数，定义如下：

```
BOOL LineTo(
    _In_ HDC hdc,
    _In_ int nXEnd,
    _In_ int nYEnd
);
```

下面是一个代码段:

```
MoveToEx(hdc, 100, 100, NULL);
LineTo(hdc, 200, 100);
LineTo(hdc, 200, 200);
LineTo(hdc, 100, 200);
LineTo(hdc, 100, 100);
```

绘制结果如图 7-13 所示。

图 7-13 绘制结果

如果只是想绘制四边形，还可以直接调用 Rectangle 函数：

```
BOOL Rectangle(
    _In_ HDC hdc,
    _In_ int nLeftRect,
    _In_ int nTopRect,
    _In_ int nRightRect,
    _In_ int nBottomRect
```

```
);
```

绘制圆或者圆弧调用 AngleArc 函数：

```
BOOL AngleArc(
    _In_ HDC      hdc,
    _In_ int      X,
    _In_ int      Y,
    _In_ DWORD dwRadius,
    _In_ FLOAT eStartAngle,
    _In_ FLOAT eSweepAngle
);
```

需要强调的是 AngleArc 函数实际绘制了一条线段和一段圆弧。线段的起点是当前 GDI 笔位置，线段的终点是通过圆心（X，Y）、半径 dwRadius，再加上 eStartAngle 计算的圆上的一个点。所以，如果只想绘制一段圆弧或者圆，首先需要将线段绘制的起点通过 MoveToEx 函数挪到圆弧绘制的起点，这个点坐标是：

X_0 = X+dwRadius*cos(eStartAngle*PI/180);

Y_0 = Y+dwRadius*sin(eStartAngle*PI/180);

下面是一段代码段用于绘制圆弧，代码如下:

```
MoveToEx(hdc, 100, 100, NULL);
AngleArc(hdc, 200, 200, 30, 0, 270);
```


绘制圆弧如图 7-14 所示。

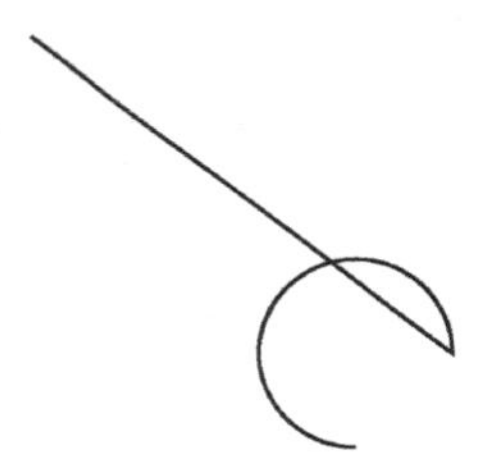

图 7-14　绘制圆弧

另一种椭圆弧（圆弧）绘制方法是使用一个矩形包围盒的 Arc 函数：

```
BOOL Arc(
    _In_ HDC hdc,
    _In_ int nLeftRect,
    _In_ int nTopRect,
    _In_ int nRightRect,
    _In_ int nBottomRect,
    _In_ int nXStartArc,
    _In_ int nYStartArc,
    _In_ int nXEndArc,
    _In_ int nYEndArc
);
```

下面是一个代码段，用于绘制椭圆弧，代码如下:

```
Rectangle(hdc, 100, 200, 450, 300);
Arc(hdc, 100, 200, 450, 300, 450, 250, 450, 250);
```

绘制椭圆弧如图 7-15 所示。

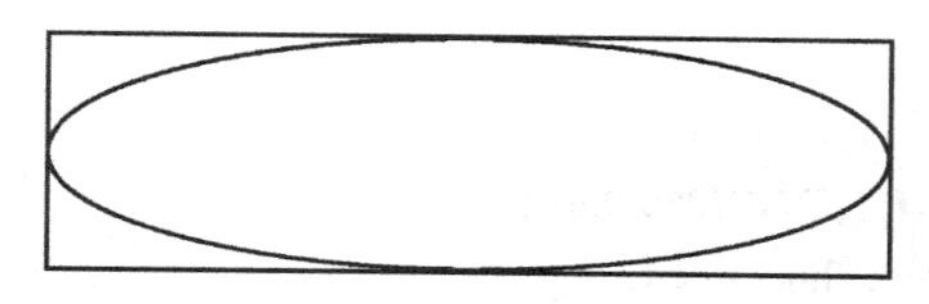

图 7-15 绘制椭圆弧

如果要绘制一个多边形，则可以调用 Polyline 函数：

```
BOOL Polyline(
    _In_        HDC     hdc,
    _In_ const POINT *lppt,
    _In_        int     cPoints
);
```

Polyline 函数传入若干有序排列的点，可用于绘制多边形，参考代码如下：

```
POINT pnts[4];
pnts[0].x = 100;
pnts[0].y = 100;
pnts[1].x= 120;
pnts[1].y = 130;
pnts[2].x = 210;
pnts[2].y = 100;
pnts[3].x = 230;
pnts[3].y = 80;
Polyline(hdc, pnts, 4);
```

绘制多边形如图 7-16 所示。

图 7-16 绘制多边形

7.4.2 笔、画刷、填充绘制

GDI 设备在绘制图形时既可以用笔也可以用画刷。CreatePen 函数的作用是创建一个笔：

```
HPEN CreatePen(
    _In_ int        fnPenStyle,
    _In_ int        nWidth,
    _In_ COLORREF crColor
);
```

其中，第一个参数设置笔的风格，常用风格是实心笔 PS_SOLID，还可以选择划线笔 PS_DASH，其他风格的笔，读者可以查阅 MSDN 网站。第二个参数定义笔的粗细度。第三个参数定义笔的颜色，这里可以使用 RGB 宏来简单地组合红、绿、蓝三种颜色。参考代码如下：

```
HPEN hPen, oldPen;
hPen = CreatePen(PS_DASH, 1, RGB(255,0,0));
oldPen = (HPEN)::SelectObject(hdc, hPen);
...
SelectObject(hdc, oldPen);
DeleteObject(hPen);
```

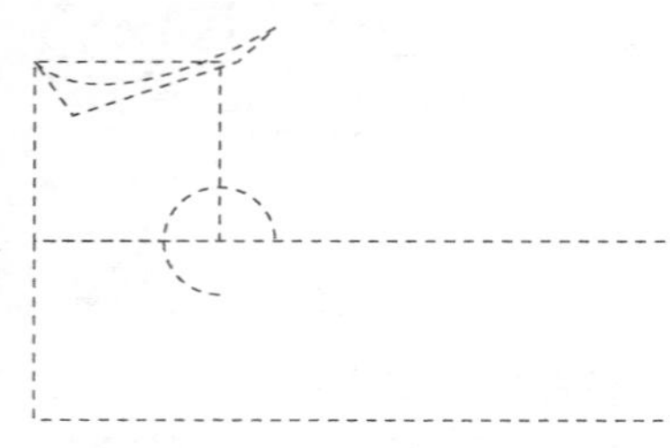

图 7-17　红色划线笔

当调用 CreatePen 函数后，它返回的是自定义笔的句柄，接着需要调用 SelectObject 函数将这只笔选到设备句柄中，SelectObject 函数返回的是一只旧的笔，一般可以保存旧的笔，当自定义的笔绘制完，再把旧的笔放回到设备句柄中。最后，一定不要忘记释放笔资源，使用 DeleteObject（hPen）函数。

红色划线笔如图 7-17 所示。

CreateSolidBrush 函数用于创建一个实心画刷，唯一的参数是笔刷颜色，定义如下：


```
HBRUSH CreateSolidBrush( _In_ COLORREF crColor);
```

因为是笔刷，所以需要调用填充类型的函数，这里可以使用填充矩形 FillRect 函数，定义如下：

```
int FillRect(
    HDC hDC,
    CONST RECT* lprc,
    HBRUSH hbr
);
```

调用 FillRect 函数代码如下：

```
HBRUSH hBrush;
hBrush = CreateSolidBrush(RGB(255,129, 0));
rect.left = 410;    rect.right = 470;
rect.top = 100;    rect.bottom = 500;
FillRect(hdc, &rect, hBrush);
DeleteObject(hBrush); //delete it
```

填充矩形如图 7-18 所示。

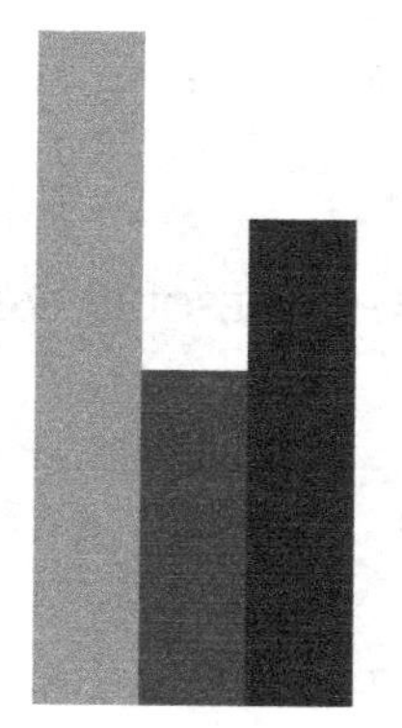

图 7-18 填充矩形

除了实心画刷外，还可以创建特殊样式的画刷：

```
HBRUSH CreateBrushIndirect(
    _In_ const LOGBRUSH *lplb
);
```

创建一个交叉风格的画刷，代码如下：

```
LOGBRUSH lplb;
lplb.lbColor=RGB(255,129, 0);
lplb.lbStyle = BS_HATCHED;
lplb.lbHatch = HS_CROSS;
hBrush = CreateBrushIndirect(&lplb);
FillRect(hdc, &rect, hBrush);
DeleteObject(hBrush); //delete it
```


不同风格的画刷如图 7-19 所示。

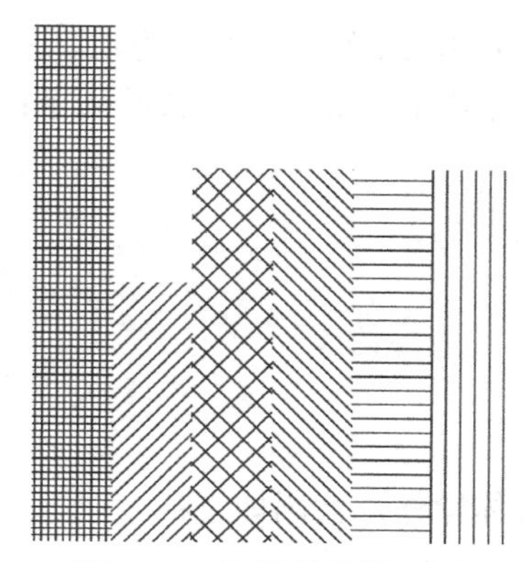

图 7-19 不同风格的画刷

除了特殊风格的画刷外，Win32 GDI 还提供位图创建画刷的函数：

```
HBRUSH CreatePatternBrush(    _In_ HBITMAP hbmp);
```

首先需要从资源中载入位图：

```
HBITMAP LoadBitmap( _In_ HINSTANCE hInstance,
    _In_ LPCTSTR      lpBitmapName);
```

调用 Create Pattern Pattern Brush 函数如下：

```
static HBITMAP hBitmap=NULL;
if(!hBitmap)
    hBitmap = LoadBitmap(hInst, MAKEINTRESOURCE(IDB_BITMAP_BURGER));
hBrush = CreatePatternBrush(hBitmap);
GetClientRect(hWnd, &rect);
FillRect(hdc, &rect, hBrush);
DeleteObject(hBrush);
```

使用位图（图像）作为画刷如图 7-20 所示。

图 7-20　使用位图（图像）作为画刷

以上使用画刷的时都采用了 FillRect 的矩形填充方法，这是一种简单的光栅复制操作。如果想使用复杂的光栅绘制，则需使用 PatBlt 方法，代码如下：

```
BOOL PatBlt(  _In_ HDC     hdc,
   _In_ int     nXLeft,
   _In_ int     nYLeft,
   _In_ int     nWidth,
   _In_ int     nHeight,
   _In_ DWORD dwRop);
```

最后一个参数就是光栅的具体操作，代码如下：

```
hBrush = CreatePatternBrush(hBitmap2);
::SelectObject(hdc, hBrush);
GetClientRect(hWnd, &rect);
PatBlt(hdc, rect.left, rect.top, rect.right-rect.left, rect.bottom-rect.top, PATCOPY);
DeleteObject(hBrush);
```

PatBlt 光栅复制的绘制方法如图 7-21 所示。

图 7-21 PatBlt 光栅复制的绘制方法

以上画刷绘制函数都是矩形绘制函数，如何绘制不规则的形状呢？GDI 提供了若干区域函数可以绘制任意形状。这里只展示两个函数 CreateEllipticRgn 和 FillRgn。

创建椭圆形区域函数 CreateEllipticRgn：

```
HRGN CreateEllipticRgn(
    _In_ int nLeftRect,
    _In_ int nTopRect,
    _In_ int nRightRect,
    _In_ int nBottomRect
);
```

用刷子填充区域函数 FillRgn：

```
BOOL FillRgn(
    _In_ HDC      hdc,
    _In_ HRGN     hrgn,
    _In_ HBRUSH hbr);
```


绘制椭圆形区域代码如下：

```
hBrush = CreatePatternBrush(hBitmap3);
::SelectObject(hdc, hBrush);
GetClientRect(hWnd, &rect);
bgRgn = CreateEllipticRgn(rect.left, rect.top, rect.right, rect.bottom);
FillRgn(hdc, bgRgn, hBrush);
DeleteObject(bgRgn);
DeleteObject(hBrush);
```

绘制椭圆形区域如图 7-22 所示。这里创建了一个椭圆形区域，然后用位图刷子绘制这个区域，最后依然不要忘记释放资源。

图 7-22 绘制椭圆形区域

7.4.3 字体和文本

这里介绍三个文本绘制函数。首先来看最简单的文本输出函数 TextOut：

```
BOOL TextOut(
    _In_ HDC        hdc,
    _In_ int        nXStart,
    _In_ int        nYStart,
    _In_ LPCTSTR lpString,
    _In_ int        cchString
);
```

其中，nXStart 和 nYStart 表示文本绘制的起始位置，lpString 要求绘制的文本字符串，cchString 要求绘制的字符串长度。

另一个文本绘制函数是 DrawText，它以某种格式进行文本绘制，定义 DrawText 函数如下：

```
int DrawText(
    HDC hDC,
    LPCTSTR lpString,
    int nCount,
    LPRECT lpRect,
    UNIT uFormat
);
```


最后一个文本绘制函数是 ExtTextOut，它允许定义绘制的裁剪区域，定义 ExtTextOut 函数如下：

```
BOOL ExtTextOut(
    HDC hdc,
    int X,
    int Y,
    UINT fuOptions,
    const RECT* lprc,
    LPCTSTR lpString,
    UINT cbCount,
    const int* lpDx
);
```

如果要设置文本颜色，使用 SetTextColor 函数：

```
COLORREF SetTextColor(
    _In_ HDC          hdc,
    _In_ COLORREF crColor
);
```

一个简单的文本绘制案例代码如下：

```
SetBkColor(hdc, RGB(200,255,255));
SetRect(&textRect, 410, 310, 800,560);
```

```
    SetTextColor(hdc, RGB(255, 0, 0));
    DrawText(hdc,TEXT("Hello, DrawText!"),-1,&textRect, DT_CENTER | DT_NOCLIP);
    TextOut(hdc,410 , 200, _T("Hello, TextOut!"), strlen("Hello, TextOut!"));
    ExtTextOut(hdc, 100, 300, ETO_OPAQUE , NULL, _T("Hello, ExtTextOut!"), strlen("Hello, ExtTextOut!"),
NULL);
```

绘制文本如图 7-23 所示。

Hello, TextOut!

Hello, ExtTextOut!

Hello, DrawText!

图 7-23 绘制文本

以上绘制函数采用设备中的默认字体，GDI 还提供了 CreateFont 函数用来创建字体，定义 CreateFont 函数如下：

```
HFONT CreateFont(
  _In_ int       nHeight,
  _In_ int       nWidth,
  _In_ int       nEscapement,
  _In_ int       nOrientation,
  _In_ int       fnWeight,
  _In_ DWORD     fdwItalic,
  _In_ DWORD     fdwUnderline,
  _In_ DWORD     fdwStrikeOut,
  _In_ DWORD     fdwCharSet,
  _In_ DWORD     fdwOutputPrecision,
  _In_ DWORD     fdwClipPrecision,
  _In_ DWORD     fdwQuality,
  _In_ DWORD     fdwPitchAndFamily,
  _In_ LPCTSTR lpszFace);
```

这是一个令人相当头痛的函数。但是，MSDN 网站中提供了每个参数的详细解释，MSDN 还提供了若干示例供程序员使用。这里给出一个简单的用法，代码如下：

```
    HFONT hFont=NULL, hOldFont=NULL;
    hFont =
  CreateFont(48,0,0,0,FW_DONTCARE,FALSE,TRUE,FALSE,DEFAULT_CHARSET,OUT_OUT
LINE_PRECIS, CLIP_DEFAULT_PRECIS,CLEARTYPE_QUALITY, VARIABLE_PITCH,TEXT("Impact"));
    hOldFont = (HFONT)SelectObject(hdc, hFont);
    DrawText(hdc,TEXT("Hello, DrawText!"),-1,&textRect, DT_CENTER );
    TextOut(hdc,410 , 200, _T("Hello, TextOut!"), strlen("Hello, TextOut!"));
    ExtTextOut(hdc, 100, 300, ETO_OPAQUE , NULL, _T("Hello, ExtTextOut!"), strlen("Hello, ExtTextOut!"),
NULL);
    SelectObject(hdc, hOldFont);
    DeleteObject(hFont);
```

创建字体如图 7-24 所示。

需要注意的是最后一个参数 lpszFace，定义字体种类。上例使用了 Impact 字体，这是一种很有冲击感的字体，如果想替换另一种字体种类，该如何定义？请参考如图 7-25 所示的字体种类，打开 Word 或者 WPS 软件，这里列出的所有字体名称都可以直接表达这个参数。

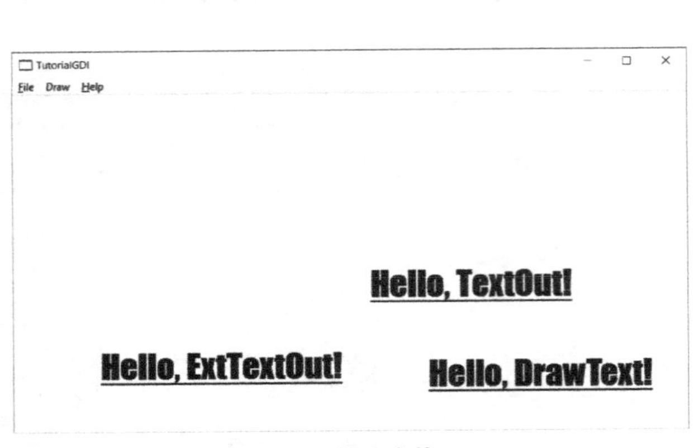

图 7-24 创建字体

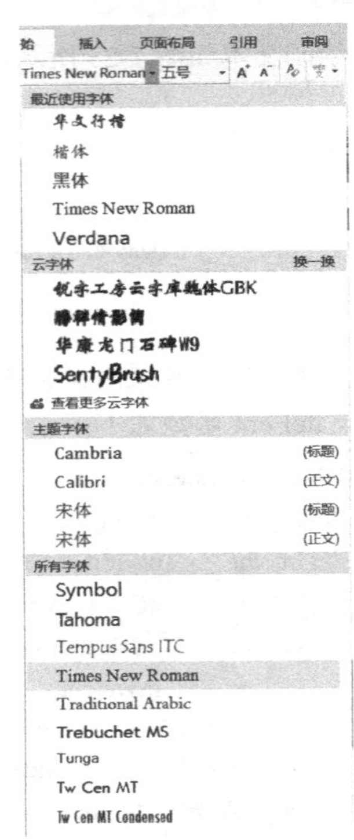

图 7-25 字体种类

7.4.4 光栅操作

光栅(Raster)是由像素点构成的一个矩形网格，每个格子就是一个像素点。光栅操作就是对这个矩形区域的操作。SetROP2 函数可以设置光栅混合模式，定义如下：

```
int SetROP2(
    _In_ HDC hdc,
    _In_ int fnDrawMode
);
```

混合模式有十几种类型，常用的有以下三种。

（1）R2_COPYPEN 复制模式：用当前画笔颜色将光栅格子中的已有的像素色进行覆盖，这也是默认的光栅模式。

（2）R2_MERGEPEN 合并模式：将当前画笔颜色和光栅格子中已有的颜色相加。

（3）R2_XORPEN 异或模式：将当前画笔颜色和光栅格子中已有的颜色异或。

图 7-26 展示了 R2_COPYPEN 的覆盖模式。首先绘制红色矩形，再绘制绿色矩形，最后绘制蓝色矩形，三次绘制形成了覆盖关系。

图 7-27 展示了 R2_MERGEPEN 的合并模式。最先绘制的背景采用黑色，黑色的颜色值

为 0。之后绘制红色矩形，再绘制绿色矩形，绿色和红色重叠的区域就是红绿的混合色：黄色。最后绘制蓝色矩形，红绿蓝重叠的区域就是白色。

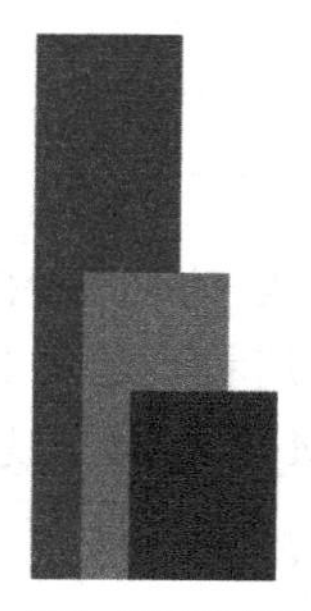

图 7-26　R2_COPYPEN 的覆盖模式

图 7-27　R2_MERGEPEN 覆盖模式

图 7-28 采用了 R2_XORPEN 的异或模式，绘制代码如下：

```
SetROP2(hdc, R2_XORPEN); //set raster operation as XOR

hPen = CreatePen(PS_SOLID, 60, RGB(255,0,0));
SelectObject(hdc, hPen);
left = 210;
rect.left = left;    rect.right = left+60; rect.top = 100;    rect.bottom = 500;
Rectangle(hdc, rect.left, rect.top, rect.right, rect.bottom);
DeleteObject(hPen);

hPen = CreatePen(PS_SOLID, 60, RGB(0,255,0));
SelectObject(hdc, hPen);
left+=40;
rect.left = left;    rect.right = left+60; rect.top = 300;    rect.bottom = 500;
Rectangle(hdc, rect.left, rect.top, rect.right, rect.bottom);
DeleteObject(hPen);

hPen = CreatePen(PS_SOLID, 60, RGB(0,0,255));
SelectObject(hdc, hPen);
left+=40;
rect.left = left;    rect.right = left+60; rect.top = 400;    rect.bottom = 500;
Rectangle(hdc, rect.left, rect.top, rect.right, rect.bottom);
DeleteObject(hPen);
```

图 7-28　R2_XORPEN 的异或模式

R2_XORPEN 模式最经典的用法就是矩形框的橡皮擦功能，感兴趣的读者可以从网上搜索该功能的具体做法。

7.4.5 双缓冲机制

为了避免绘制闪烁问题，需要使用双缓冲：即在后台缓冲(Back Buffer) 中绘制，然后将绘制的后台缓冲直接刷新到前台缓冲(Front Buffer)。前台缓冲即用户看到的内容。双缓冲代码具有固定的套路，读者不需要记忆，只要将类似的代码复制到 WM_PAINT 消息中即可，方法如下。

（1）定义设备描述表及位图句柄。

```
HDC hMemDC;
HBITMAP hBitmap;
```

（2）创建一个与窗口矩形显示兼容的内存显示设备描述表。

```
hMemDC= CreateCompatibleDC(hDC);
```

（3）用 hDC 创建一个与窗口矩形显示兼容的位图。

```
hBitmap= CreateCompatibleBitmap(hDC, rt.right - rt.left, rt.bottom -rt.top);
```

（4）将位图 hBitmap 选到内存显示设备 hMemDC 中，只有选入位图的内存显示设备才有地方绘图，画到指定的位图上。


```
SelectObject(hMemDC,hBitmap);
```

（5）用 hDC 背景色将位图清除干净。

```
FillRect(hMemDC,&rt, hBrush)
```

（6）绘图。

（7）将内存中的图复制到窗口矩形上显示。

```
BitBlt(hdc, 0, 0, rt.right - rt.left,rt.bottom - rt.top, hMemDC, 0, 0, SRCCOPY)
```

使用双缓冲示例参考代码如下：

```
case WM_PAINT:
    hDC = BeginPaint(hWnd, &ps);
    GetClientRect(hWnd,&clientRect);
    hMemDC = CreateCompatibleDC(hDC);
    hBmp = CreateCompatibleBitmap(hDC, clientRect.right, clientRect.bottom);
    SelectObject(hMemDC,hBmp);
    FillRect(hMemDC,&clientRect, GetSysColorBrush(COLOR_3DFACE));
    // 在这里绘制你自己的内容 …..

    BitBlt(hDC,0,0,clientRect.right,clientRect.bottom,hMemDC,0,0,SRCCOPY);
    DeleteObject(hBmp);
    DeleteDC(hMemDC);
    EndPaint(hWnd, &ps);
```

```
break;
```

另外需要注意在注册窗口类的时候还需要将默认背景刷子设置为 NULL，方法是在 MyRegisterClass 中设置 wcex.hbrBackground=NULL；当响应 WM_PAINT 消息时，BeginPaint 不使用刷子重新刷新 DC。

第 8 章　综合编程实例

8.1　扑克洗牌

给定 *N* 副扑克，每副扑克有 54 张牌，请写出扑克牌的洗牌程序，并打印洗牌结果。要求将结果存储到文件中，并能够再次读取。程序要求：扑克牌中的 4 种花色及大王、小王用一个枚举类型表示，扑克牌 1～13 用 short 类型表示，每张牌用一个结构体表示。

扑克是桌游中非常重要的一个分支，定义牌及洗牌是任何扑克游戏必有的功能。这并不是一道竞赛题，但是该题涉及枚举、结构、数组等基本数据类型及文件 I/O、随机数等，涉及知识面较广。

先来看扑克牌花色的定义。题目要求使用枚举类型，读者一定要避免使用汉语拼音定义枚举成员，大家可以从维基百科中获得扑克牌中花色的各种英文单词，链接地址为 https://en.wikipedia.org/wiki/Suit_(cards)。题目要求打印扑克，怎样打印花色呢？在 Console 窗口下读者可以尝试如下代码：

```
for(int i=1; i<=6; ++i)
    cout<<i<<" "<<(char)i<<"    ";
```

在 Windows 操作系统下（非 Win10 操作系统），扑克牌花色打印结果如图 8-1 所示。

图 8-1　扑克牌花色打印结果

不难看出，可以用 ASCII 码的 1 表示小王，2 表示大王，3 表示红桃，4 表示方块，5 表示梅花，6 表示黑桃。因此不难写出如下花色枚举类型的定义，代码如下：

```
enum Suit {BLACK_JOKER=1, RED_JOKER, HEART, DIAMOND, CLUB, SPADE};
```

定义了花色枚举类型，不难定义每张牌的结构体，代码如下：

```
struct Deck
{
    Suit suit; //花色
    short num; //大小：1 (Ace), 2, 3, 4, 5, 6, 7, 8, 9, 10, 11(Jack), 12(Queen), 13(King)
```

```
};
```

以下代码是 *N* 副扑克的定义，这里使用类 Poker 进行封装，它包含了牌的初始化、洗牌、打印、保存与载入。数据成员 m_PackNum 表示该游戏使用几副牌，m_Decks 是所有牌的存储数组。Poker 类声明如下，代码如下：

```
class Poker
{
    public:
        //构造函数，初始化几副牌
        Poker(int packNum);
        ~Poker();
        //自定义复制构造函数，避免浅复制问题
        Poker(const Poker& poker);
        //重载操作符等于号，避免浅复制问题
        Poker& operator=(const Poker& poker);
        //初始化扑克牌
        void Init();
        //洗牌
        void Shuffle();
        //保存当前牌
        bool Save(const char* filename);
        //载入扑克牌
        bool Load(const char* filename);
        //打印扑克牌
        void Print();
    private:
        int m_PackNum;    //几副牌
        Deck* m_Decks;    //所有牌都存储在这个数组中
};
```

因为所有扑克牌存储在一个动态数组中，所以要特别注意指针深复制与浅复制问题，这里自定义了复制构造函数和操作符等于号函数，代码如下：

```
//构造函数
Poker::Poker(int packNum)
{
    m_PackNum = packNum;
    m_Decks = new Deck[m_PackNum*54];
    Init();
}
//析构函数，要释放扑克牌内存
Poker::~Poker()
{
    delete[] m_Decks;
```

```
}
//自定义复制构造函数，避免浅复制问题
Poker::Poker(const Poker& poker)
{
    m_PackNum = poker.m_PackNum;
    m_Decks = new Deck[m_PackNum*54];
    memcpy(m_Decks, poker.m_Decks, sizeof(Deck)*m_PackNum*54);
}
//自定义操作符等于号函数，避免浅复制问题
Poker& Poker::operator=(const Poker& poker)
{
    if(&poker= =this)
        return *this;
  //先释放自身分配的动态内容
delete[] m_Decks;
    m_PackNum = poker.m_PackNum;
    m_Decks = new Deck[m_PackNum*54];
    memcpy(m_Decks, poker.m_Decks, sizeof(Deck)*m_PackNum*54);
    return *this;
}
```


一副新扑克牌，所有牌按照花色及大小顺序存放。这里提供一个函数 Init 定义新扑克牌的顺序，代码如下：

```
//初始化 N 副扑克
void Poker::Init()
{
    int i, j, id;
    for(i=0; i<m_PackNum; ++i)
    {
        id = i*54;
        for(j=0; j<13; ++j)
        {
            m_Decks[id+j].suit = SPADE;
            m_Decks[id+j].num = j+1;
            m_Decks[id+j+13].suit = HEART;
            m_Decks[id+j+13].num = j+1;
            m_Decks[id+j+26].suit = DIAMOND;
            m_Decks[id+j+26].num = j+1;
            m_Decks[id+j+39].suit = CLUB;
            m_Decks[id+j+39].num = j+1;
        }
        m_Decks[id+52].suit = BLACK_JOKER;
        m_Decks[id+52].num = 0; //大小王的牌值设置为 0
        m_Decks[id+53].suit = RED_JOKER;
```

```
            m_Decks[id+53].num = 0; //大小王的牌值设置为 0
        }
    }

    //打印扑克牌
    void Poker::Print()
    {
        int i, j, id;
        for(i=0; i<m_PackNum; ++i)
        {
            id = i*54;
            for(j=0; j<54; ++j)
            {
                std::cout<<(char)m_Decks[id+j].suit<<" "<<std::setw(2)<<m_Decks[id+j].num<<"    ";
                if(j!=0 && (j+1)%13==0)
                    std::cout<<"\n";
            }
            std::cout<<"\n";
        }
    }
```


定义洗牌函数 Shuffle，首先使用当前时间作为 srand 函数的随机数种子，然后对当前循环中的每张牌都生成一个随机数，让当前牌和该随机数指向的牌进行交换，代码如下：

```
    void Poker::Shuffle()
    {
        int i, num, j;
        srand(time(NULL));
        num = m_PackNum*54;
        for(i=0; i<num; ++i)
        {
            j = rand()%num;
            std::swap(m_Decks[i], m_Decks[j]);
        }
    }
```

扑克的保存与载入都采用了二进制文件方式。因为相比于文本文件，二进制文件保存更简单，可以直接将所有牌的内容一次性写入或读取。在读取扑克时，需要注意可能读取失败的问题，因此需要处理各种异常情况，代码如下：

```
    bool Poker::Save(const char* filename)
    {
        std::ofstream file(filename, std::ios::binary);
        if(file.fail())
            return false;
        //先写出几副牌数
```

```
file.write((char*)(&m_PackNum), sizeof(m_PackNum));
    //再写出所有牌数据
file.write((char*)m_Decks, sizeof(Deck)*m_PackNum*54);
    file.close();
    return true;
}
//载入扑克牌
bool Poker::Load(const char* filename)
{
    std::ifstream file(filename, std::ios::binary);
    if(file.fail())
        return false;
    int packNum;
    Deck *decks=NULL;
    //先读入几副牌数
file.read((char*)(&packNum), sizeof(packNum));
    if(file.fail() || packNum<=0)
    {
        file.close();
        return false;
    }
    //分配足够的内存
    decks = new Deck[packNum*54];
    //读入所有的牌数据
file.read((char*)decks, sizeof(Deck)*packNum*54);
    if(file.fail())
    {
        file.close();
        delete[] decks;
        return false;
    }
    delete[] m_Decks;
    m_Decks = decks;
    m_PackNum = packNum;
    file.close();
    return true;
}
```

8.2 二叉树重建可视化

第 4.1.6 节给出后序遍历和中序遍历重构一棵二叉树的算法。为了验证算法的正确性，将二叉树结果可视化则更完美。第 7 章已经了解 Win32 消息机制及 GDI 绘制的各种方法，下面就以二叉树重建可视化为例，将算法和图形绘制结合进行综合程序设计。二叉树绘制如图 8-2 所示，这也是程序最终界面。菜单 File 下设置 Open 菜单项，用户单击 Open 后会打开 binarytree.dat 文件，该文件第一行是一棵二叉树后序遍历的结果，第二行是一棵二叉树中序遍历的结果，之后就绘制如图 8-2 所示的二叉树。

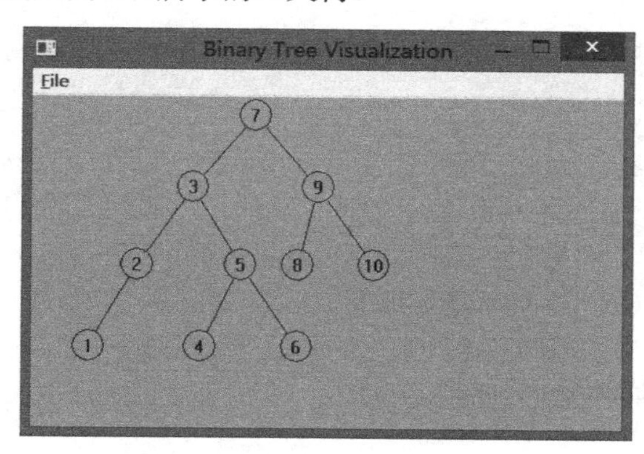

图 8-2　二叉树

对于一颗二叉树，需要绘制三部分内容：结点、分支和值。结点就是图 8-2 中的圆，分支就是图 8-2 中的线段，值就是其中的整数文字。

这里在第 4.1.6 节综合编程实例——构建二叉树代码基础上进行修改，首先是树结点。树结点中添加三个结构成员变量 x, y 和 spanAngle。这三个参数都是为绘制准备的，他们分别表示绘制时结点的 x 参考坐标，结点的 y 参考坐标，以及从当前结点绘制到两个子结点时的张角。需要指出的是，这里的参考坐标是指当绘制窗口大小发生变化时，用参考坐标计算实际的二维 GDI 坐标，代码如下：


```
struct TreeNode
{
    TreeNode()
    {
        left = right = NULL;
        value = 0;
        x = y = 0;
        spanAngle = 30*PI/180;
    }
    int value;          //结点值
    TreeNode *left; //左儿子
    TreeNode *right;//右儿子
    //以下三个参数主要为了绘制结点
    double x;           //绘制时该结点的参考 X 坐标
    double y;           //绘制时该结点的参考 Y 坐标
    double spanAngle; //绘制时该结点的张角
};
```

为绘制一棵完整的二叉树，这里封装了类 TreeDrawer。类成员包含一棵二叉树 m_pRoot，二叉树实际包围盒 m_BoundBox，结点半径 m_NodeRadius，分支长度 m_BranchLength，初始张角 m_SpanAngle，从实际包围盒缩放到绘制窗口的缩放比例 m_Scale。TreeDrawer 类声明如下：

```
//二叉树绘制器
class TreeDrawer
{
public:
    //构造函数传入一棵二叉树
    TreeDrawer(const TreeNode* pRoot=NULL);
    //初始化树结点
    void InitTreeNode(const TreeNode* pRoot);
    //调整初始张角
    void AdjustSpanAngle(bool bSpan);
    //调整分支长度
    void AdjustBranchLength(bool bAdd);
    //调整结点半径
    void AdjustNodeRadius(bool bAdd);
    //刷新包围盒，每次初始化一个新二叉树的时候一定要刷新包围盒
    void RefreshBoundBox();
    //绘制二叉树
    void Draw(RECT clientRect, HDC hdc);

private:
    //二叉树根结点
    TreeNode* m_pRoot;
    //二叉树的实际包围盒，在绘制时该包围盒内容将被平移及缩放到实际窗口中
    RECT m_BoundBox;
    //结点半径
    double m_NodeRadius;
    //分支长度
    double m_BranchLength;
    //初始张角
    double m_SpanAngle;
    //从实际包围盒缩放到绘制窗口的缩放比例
    double m_Scale;
};
```

TreeDrawer 构造函数如下：

```
TreeDrawer::TreeDrawer(const TreeNode* pRoot)
{
    m_pRoot = const_cast<TreeNode*>(pRoot);
```

```
        m_NodeRadius = 10;
        m_BranchLength = 40;
        m_SpanAngle = 50*PI/180;
        m_Scale = 1.0;
        RefreshBoundBox();
    }
```

TreeDrawer 函数初始化树如下：

```
    void TreeDrawer::InitTreeNode(const TreeNode* pRoot)
    {
        m_pRoot = const_cast<TreeNode*>(pRoot);
        RefreshBoundBox();
    }
```

TreeDrawer 函数中一个重要成员函数是 RefreshBoundBox，它不仅要刷新包围盒，还要计算每个结点的位置、张角。另外，一个很重要的功能是避免结点之间的碰撞。这里采用宽度优先遍历方法计算每一层从左往右的每个结点的绘制信息，代码如下：

```
    void TreeDrawer::RefreshBoundBox()
    {
        m_BoundBox.left = 1000000;
        m_BoundBox.right = -1000000;
        m_BoundBox.top = 1000000;
        m_BoundBox.bottom = -1000000;
        if(!m_pRoot)
            return;
        std::deque<TreeNode*> seedNodes;
        std::vector<TreeNode*> levelNodes;
        TreeNode *node, *node2;
        double cosAngle, sinAngle;
        m_pRoot->x = 0;
        m_pRoot->y = 0;
        m_pRoot->spanAngle = m_SpanAngle;
        //宽度优先遍历
        seedNodes.push_back(m_pRoot);
        while(!seedNodes.empty())
        {
            node = seedNodes.front();
            seedNodes.pop_front();
            //碰撞检测
            if(levelNodes.size())
            {
                //两个结点表示的圆是否相交
                if(levelNodes.back()->x+m_NodeRadius > node->x-m_NodeRadius &&
                    fabs(levelNodes.back()->y-node->y)<10)
```

```
                {
                    //简单避障法：右边的结点往右移动，避免碰撞
                    node->x=levelNodes.back()->x
+m_NodeRadius+m_NodeRadius*1.2;
                }
            }
            levelNodes.push_back(node);
            //计算包围盒
            if(node->x<m_BoundBox.left)
                m_BoundBox.left = node->x;
            if(node->x>m_BoundBox.right)
                m_BoundBox.right = node->x;
            if(node->y<m_BoundBox.top)
                m_BoundBox.top = node->y;
            if(node->y>m_BoundBox.bottom)
                m_BoundBox.bottom = node->y;
            cosAngle = cos(node->spanAngle);
            sinAngle = sin(node->spanAngle);
            //计算左儿子结点
            if(node->left)
            {
                node2 = node->left;
                //计算子结点圆心位置
                node2->x = node->x-cosAngle*(m_BranchLength+2*m_NodeRadius);
                node2->y = node->y+sinAngle*(m_BranchLength+2*m_NodeRadius);
                //每降一层，分支张角都变大（实际绘制是变小）
                node2->spanAngle = node->spanAngle+PI*5/180;
                if(node2->spanAngle>80*PI/180)
                    node2->spanAngle = 80*PI/180;
                seedNodes.push_back(node2);
            }
            //计算右儿子结点
            if(node->right)
            {
                node2 = node->right;
                //计算子结点圆心位置
                node2->x = node->x+cosAngle*(m_BranchLength+2*m_NodeRadius);
                node2->y = node->y+sinAngle*(m_BranchLength+2*m_NodeRadius);
                //每降一层，分支张角都变大（实际绘制是变小）
                node2->spanAngle = node->spanAngle+PI*5/180;
                if(node2->spanAngle>80*PI/180)
                    node2->spanAngle = 80*PI/180;
                seedNodes.push_back(node2);
```

```
            }
        }
    }
```

TreeDrawer 的绘制函数 Draw 也采用了宽度优先遍历。需要指出的是 RefreshBoundBox 函数虽然计算每个结点的绘制位置，但在实际窗口中，需要将结点位置做坐标映射，使其尽量铺满实际窗口。这里的坐标映射实际上就是先缩放再平移，代码如下：

```
void TreeDrawer::Draw(RECT clientRect, HDC hdc)
{
    if(!m_pRoot)
        return;
    //计算横向缩放比例
double scaleX =
(clientRect.right-clientRect.left)*0.7/(m_BoundBox.right-m_BoundBox.left);
    //计算纵向缩放比例
    double  scaleY  =
(clientRect.bottom-clientRect.top)*0.7/(m_BoundBox.bottom-m_BoundBox.top);
    //横向偏移量
    int  startX  =
clientRect.left + (clientRect.right-clientRect.left)*0.1;
    //计算整体缩放比例
    m_Scale=  std::min(scaleX, scaleY);
    //不要让缩放比例太大
    if(m_NodeRadius*m_Scale>50)
        m_Scale = 50/m_NodeRadius;
    //纵向偏移量
    int startY = m_NodeRadius*m_Scale*1.2;
    char buffer[64];


    std::deque<TreeNode*> seedNodes;
    TreeNode *node, *node2;
    double dx, dy;
    double x, y, x2, y2, offsetScale;
    SIZE size;

    //设置文字透明
    SetBkMode( hdc, TRANSPARENT );

    //宽度优先遍历
    seedNodes.push_back(m_pRoot);
    while(!seedNodes.empty())
    {
        node = seedNodes.front();
```

```
seedNodes.pop_front();
//坐标映射：缩放+平移
x = startX+(node->x-m_BoundBox.left)*m_Scale;
y = startY+(node->y-m_BoundBox.top)*m_Scale;
//绘制圆形结点
Arc(hdc, x-m_NodeRadius*m_Scale, //left
     y-m_NodeRadius*m_Scale,        //top
     x+m_NodeRadius*m_Scale,        //right
     y+m_NodeRadius*m_Scale,        //bottom
     x-m_NodeRadius*m_Scale,        //start point X
     y-m_NodeRadius*m_Scale,        //start point Y
     x-m_NodeRadius*m_Scale,        //end point X
     y-m_NodeRadius*m_Scale);       //end point Y
sprintf(buffer, "%d", node->value);
//计算需要绘制的结点值字符串的 GDI 宽度和高度
GetTextExtentPoint32(hdc, buffer, strlen(buffer), &size);
//然后让该字符串绘制在结点中心
ExtTextOut(hdc, x-size.cx*0.5, y-size.cy*0.5, ETO_OPAQUE,NULL, buffer, strlen(buffer), NULL);
if(node->left)
{
     node2 = node->left;
     x2 = startX+(node2->x-m_BoundBox.left)*m_Scale;
     y2 = startY+(node2->y-m_BoundBox.top)*m_Scale;
     dx = x2-x;
     dy = y2-y;
     offsetScale = (m_NodeRadius*m_Scale)/sqrt(dx*dx+dy*dy);
     //dx，dy 是为了让分支的起点和终点绘制在结点圆的边缘而不是圆心
     dx*=offsetScale;
     dy*=offsetScale;
     //可以将 dx, dy 设置为 0，观察变化

     seedNodes.push_back(node2);
     //绘制分支
     ::MoveToEx(hdc, x+dx, y+dy, NULL);
     ::LineTo(hdc, x2-dx, y2-dy);
}
if(node->right)
{
     node2 = node->right;
     x2 = startX+(node2->x-m_BoundBox.left)*m_Scale;
     y2 = startY+(node2->y-m_BoundBox.top)*m_Scale;
     dx = x2-x;
     dy = y2-y;
```

```
                offsetScale = (m_NodeRadius*m_Scale)/sqrt(dx*dx+dy*dy);
                //dx，dy 是为了让分支的起点和终点绘制在结点圆的边缘而不是圆心
                dx*=offsetScale;
                dy*=offsetScale;
                //可以将 dx, dy 设置为 0，观察变化

                seedNodes.push_back(node2);
                ::MoveToEx(hdc, x+dx, y+dy, NULL);
                ::LineTo(hdc, x2-dx, y2-dy);
            }
        }
    }
```

利用后序遍历和中序遍历构建一棵二叉树代码如下：

```
void BuildTree(int * postOrderStr, int* inOrderStr, int n, TreeNode *&root)
{
    if(n<=0)
        return;

    //创建当前子树的根结点
    root = new TreeNode();
    root->value = postOrderStr[n-1];

    //在中序遍历中找到后序遍历的最后一个值也就是根结点值， pos 既是左右子树的分割位置，也是左子树结点个数
    int pos = std::find(inOrderStr, inOrderStr+n, postOrderStr[n-1]) - inOrderStr;
    //后序遍历的最后一个值就是根结点值，也就是前序遍历第一个值

    //遍历构建左子树，所以要传入当前子树的左儿子结点
    BuildTree(postOrderStr, inOrderStr, pos, root->left);
    //遍历构建右子树，所以要传入当前子树的右儿子结点
    BuildTree(postOrderStr+pos, inOrderStr+pos+1, n-pos-1, root->right);
}
```

从文件中读取后序遍历和中序遍历，然后构建一棵二叉树，代码如下：

```
bool ReadTraverse(const char* filename, TreeNode *&root)
{
    int n1, n2, i;
    int inOrderStr[N], postOrderStr[N];
    std::string str;
    istringstream iss;
    std::ifstream file;

    root=NULL;
```

```
    //打开文件
    file.open(filename);
    if(file.fail())
        return false;

    //得到一行输入：后序遍历结果
    std::getline(file, str);
    //打包这行字符串到字符串流
    iss.str(str);
    n1 = 0;
    while(iss>>i)
        postOrderStr[n1++]=i;

    //得到另一行输入：中序遍历结果
    std::getline(file, str);
    //先清除标志位
    iss.clear();
    //打包这行字符串到字符串流
    iss.str(str);
    n2 = 0;
    while(iss>>i)
        inOrderStr[n2++]=i;

    //关闭文件
    file.close();
    //断言 n1==n2
    assert(n1==n2);
    if(n1!=n2)
        return false;
    BuildTree(postOrderStr, inOrderStr, n1, root);
    return true;
}
```


最后就是主窗口的消息回调函数，在此之前先定义两个全局变量 g_TreeRoot 和 g_TreeDrawer，一个是二叉树，另一个是二叉树绘制器，代码如下：

```
TreeNode *g_TreeRoot=NULL;
TreeDrawer g_TreeDrawer;
...
LRESULT CALLBACK WindowProcedure (HWND hwnd, UINT message, WPARAM wParam, LPARAM
lParam)
{
    HDC hdc;
    PAINTSTRUCT ps;
    RECT rect;
```

```
switch (message)                    /* handle the messages */
{
case WM_CREATE:
    //创建主菜单
    CreateMainMenu(hwnd);
    break;
case WM_COMMAND:
{
    //各种菜单项消息
    switch(wParam)
    {
    case IDM_FILE_OPEN:
        //先释放前一次的二叉树内存
        FreeTree(g_TreeRoot);
        g_TreeRoot = NULL;
        //再读入文件
        ReadTraverse("binarytree.dat", g_TreeRoot);
        g_TreeDrawer.InitTreeNode(g_TreeRoot);
        InvalidateRect(hwnd, NULL, 1);
        break;
    case IDM_FILE_QUIT:
        //程序退出
        DestroyWindow(hwnd);
        break;
    default:
        break;
    }
}
case WM_SIZE:
    //文档窗口的大小发生变化时，需要重新绘制整个窗口
    InvalidateRect(hwnd, NULL, 1);
    break;
case WM_PAINT:
    hdc = BeginPaint(hwnd, &ps);
    //得到窗口矩形大小
    GetClientRect(hwnd, &rect);
    //绘制二叉树
    g_TreeDrawer.Draw(rect, hdc);
    EndPaint(hwnd, &ps);
    break;
case WM_DESTROY:
    PostQuitMessage (0);        /* send a WM_QUIT to the message queue */
    break;
```

```
        default:                              /* for messages that we don't deal with */
            return DefWindowProc (hwnd, message, wParam, lParam);
        }
        return 0;
    }
```

对于稍微复杂的二叉树，简单的碰撞检测和避障效果如图 8-3 所示。读者可以修改避障代码使左边的结点往左挪一些，从而改善树的可视化效果。

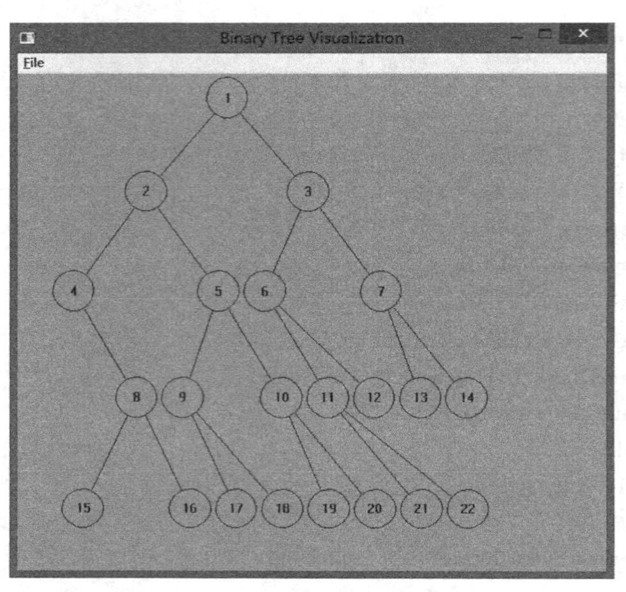

图 8-3　简单的碰撞检测和避障效果

8.3　L-System 分形树建模

L-System 是 Lindenmayer System 的简称，这是一套基于规则语法的字符串重写系统。一个 L-System 包含一套语法符号和对应的生成规则，这套语法符号中包含一个初始的原子字符串，通过对应的生成规则，最终将原子字符串重写扩充成一个复杂的字符串。而这个最终的字符串就是一个复杂的几何结构。

L-System 有多种表达形式，例如，参数 L-System，约束 L-System，上下文相关 L-System 等。这里介绍最常用和最简单的参数 L-System，它以三元组的形式定义如下：

$$G=(V, \omega, P)$$

其中，V 是一组符号，它包含一组变量和常量；ω（原子符号或初始串）是一个原子字符串，又称为初始字符串；P 是一组生成规则，它定义变量的演化生成规则。

1. 经典 L-System 案例：海藻

Lindenmayer 最原始的 L-System 案例就是海藻生长模拟系统，定义如下：

变量：A B

常量：无

原子符号：A

生成规则：(A → AB), (B → A)

每次迭代后（n 表示迭代次数），产生的字符串如下:

$n=0$: A
$n=1$: AB
$n=2$: ABA
$n=3$: ABAAB
$n=4$: ABAABABA
$n=5$: ABAABABAABAAB
$n=6$: ABAABABAABAABABAABABA
$n=7$: ABAABABAABAABABAABABAABAABABAABAAB

迭代过程如下：

```
n=0:            A              起始(原子符号串)
               / \
n=1:          A   B            A 通过规则 A→AB 演化成了 AB
             / \   \
n=2:        A   B   A          A 通过规则 A→AB 演化成 AB，B 通过规则 B→A 演化成 A
           / \  |   /\
n=3:      A   B A  A  B
         / \  | /\ /\  \
n=4:    A  BA ABAB  A
```

一个有趣的现象是，这些字符串的长度是一个 Fibonacci 数组。

1 2 3 5 8 13 21 34 55 89 ...

2. 经典 L-System 案例：简单分形树

变量: 0, 1

常量: [,]

原子符号 : 0

生成规则 : (1 → 11), (0 → 1[0]0)

根据以上规则可以得到:

原子符号: 0
1 次迭代: 1[0]0
2 次迭代: 11[1[0]0]1[0]0
3 次迭代: 1111[11[1[0]0]1[0]0]11[1[0]0]1[0]0
...

可以看到这组字符串快速增长，利用海龟作图方法可以对该字符串绘制一幅图。例如，可以定义如下绘图规则。

0：绘制一条线段。

1：绘制一条线段。

[：把当前位置和角度压入栈，然后左转 45 度。

]：把当前位置和角度从栈中弹出，然后右转 45 度。

根据如上规则，可以得到分形树，简单分形树的演变如图 8-4 所示。

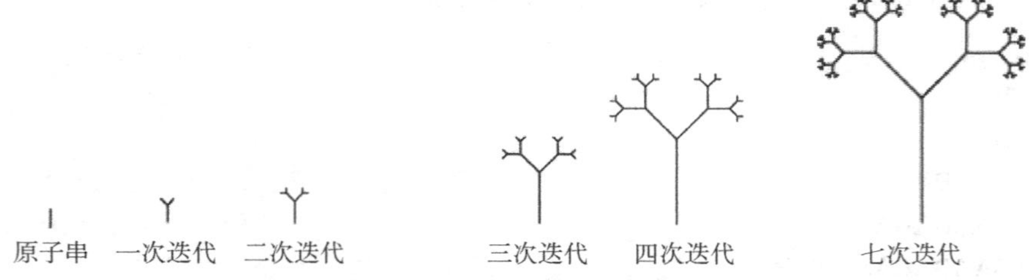

图 8-4　简单分形树的演变

3. 经典 L-System 案例：康托尔集合

变量：A B

常量：无

原子符号：A

生成规则：（A→ABA），（B→BBB）

这里 A 代表向前绘制一段距离，而 B 代表向前移动一段距离，从而生成著名的康托尔分形集合。康托尔集合如图 8-5 所示。

图 8-5　康托尔集合

需要注意的是，如果读者想绘制如图 8-5 所示的康托尔集合，需要将每次迭代的单位距离缩小到上一次迭代距离的三分之一。

4. 经典 L-System 案例：Koch 曲线

变量：F

常量：+ −

原子符号：F

生成规则：（F→F+F−F−F+F）

这里，F 代表向前绘制一段距离，+代表左转 90°，− 代表右转 90°。

```
n = 0:    F
n = 1:    F+F−F−F+F
n = 2:    F+F−F−F+F + F+F−F−F+F − F+F−F−F+F − F+F−F−F+F + F+F−F−F+F
n = 3:    F+F−F−F+F+F+F−F−F+F−F+F−F−F+F−F+F−F−F+F+F+F−F−F+F +
F+F−F−F+F+F+F−F−F+F−F+F−F−F+F−F+F−F−F+F+F+F−F−F+F −
F+F−F−F+F+F+F−F−F+F−F+F−F−F+F−F+F−F−F+F+F+F−F−F+F −
F+F−F−F+F+F+F−F−F+F−F+F−F−F+F−F+F−F−F+F+F+F−F−F+F +
F+F−F−F+F+F+F−F−F+F−F+F−F−F+F−F+F−F−F+F+F+F−F−F+F
```

Koch 曲线如图 8-6 所示。

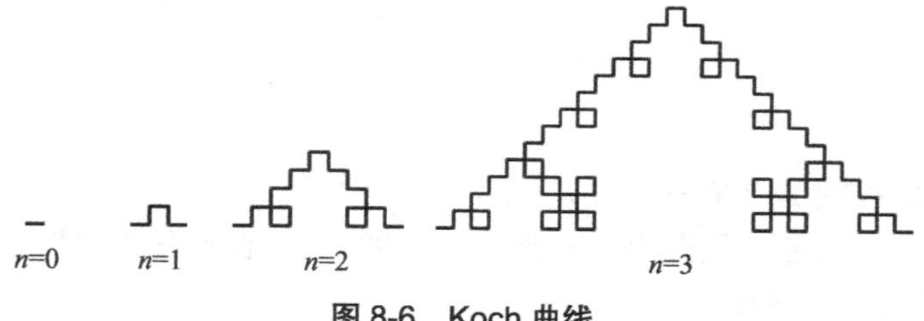

图 8-6 Koch 曲线

5. 经典 L-System 案例：Sierpinski 三角

变量：F G

常量：+ −

原子符号：F−G−G

生成规则：(F→F−G+F+G−F)，(G → GG)

这里，F 和 G 都表示向前绘制一段距离，+表示左转 120°，−表示右转 120°。

Sierpinski 三角如图 8-7 所示。

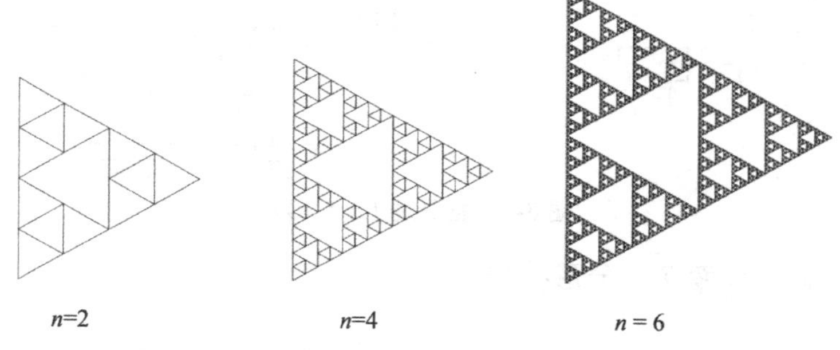

图 8-7 Sierpinski 三角

Sierpinski 三角的另一个变种。

变量：A B

常量：+ −

原子符号：A

生成规则：(A → +B−A−B+)，(B → −A+B+A−)

这里 A 和 B 表示向前绘制一段距离，+表示左转 60°，而−表示右转 60°。

Sierpinski 三角变种如图 8-8 所示。

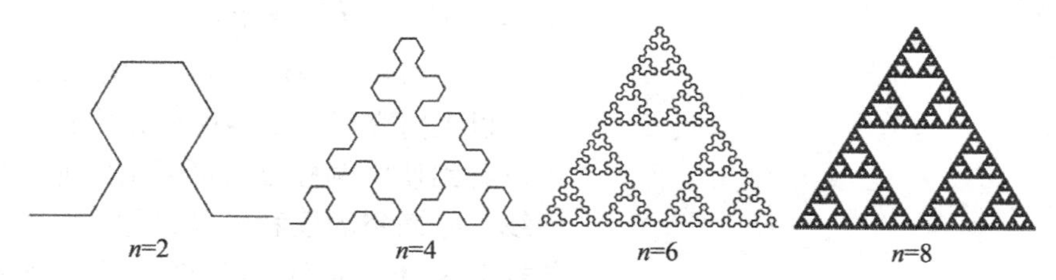

图 8-8 Sierpinski 三角变种

6. 经典 L-System 案例：龙曲线

变量：X Y

常量：F + －

原子符号：FX

规则：(X→ X+YF+)，(Y → －FX－Y)

这里，F 表示向前绘制一段距离，－表示左转 90°， +表示右转 90° 。这里 X 和 Y 都和绘制无关，只是用来控制曲线演化。

龙曲线（n=10）如图 8-9 所示。

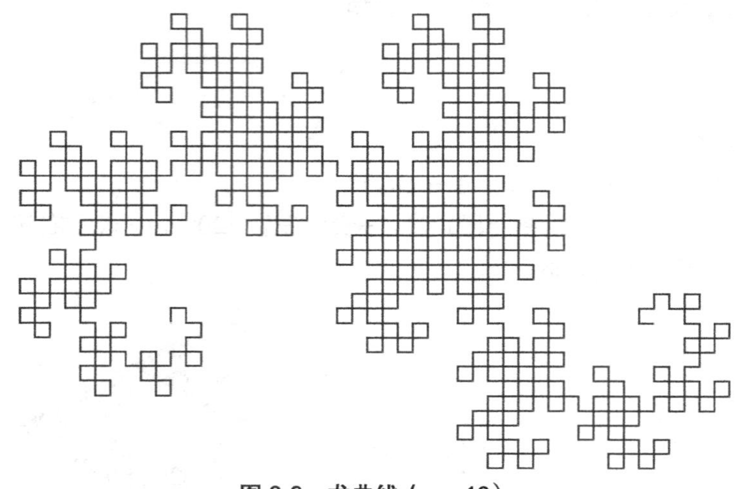

图 8-9 龙曲线（n = 10）

7. 经典 L-System 案例：分形植物

变量：X F

常量：+ － [ ]

图 8-10 分形植物（n = 8）

原子符号：X

生成规则：(X → F[－X][X]F[－X]+FX)，(F → FF)

这里，F 表示向前绘制一段距离，−表示左转 25°，+ 表示右转 25°。X 变量和绘制内容无关，只是用来控制曲线演化。"[" 表示将当前位置和角度压入堆栈，"]" 表示将当前位置和角度设置为栈顶元素值，并将栈顶弹出。

分形植物（n=8）如图 8-10 所示。

以上内容展示了各种经典的 L-System 例子，可以看到使用编程方法可以生成非常优美的分形曲线，感兴趣的读者可以自行编写程序绘制上述各种 L-System 曲线。具体细节请参考维基百科中 L-System 介绍，链接地址：https://en.wikipedia.org/

wiki/L-system。

这里以分形植物为例展示如何解析 L-System 并将其绘制成树。

编写图形程序的时候，应当遵循数据与应用框架分离的原则。数据和算法放在独立的 FractalTreeModeller.h 和 FractalTreeModeller.cpp 文件中。主程序和界面代码存放在 main.cpp 文件中。

因为分形植物建模需要用到堆栈，那么第一个问题就是堆栈中的数据类型是什么？答案是位置和角度。为了加快绘制，这里将二维平面笛卡儿坐标系下的角度用方向向量代替。堆栈中数据类型 ParaInfor 定义如下：

```
struct ParaInfor
{
    double x;        //二维位置向量
    double y;
    double dirX;     //二维方向向量
    double dirY;
};
```

接着，需要将分形植物的 L-System 程序及各种参数封装成一个类 FractalTree。FractalTree 类最重要的功能就是给定一个迭代次数 n，生成对应的符号字符串，这是由类成员函数 GenerateString 完成的。有了符号字符串后，就要将该字符串绘制成图形，这是由类成员函数 Draw 完成的。当然，FractalTree 类还提供了迭代次数、转向角度、枝干长度等参数的设置。综合上述内容，FractalTree 类声明如下：

```
class FractalTree
{
public:
    FractalTree(double baseLength=30, int n = 1, double angle = 25, double initDirAngle=30);
    ~FractalTree();
    //对外接口，生成符号串
    void GenerateString();
    //迭代次数
    void SetN(int n);
    int GetN();
    //每次枝干生长的长度
    void SetBaseLength(double len);
    double GetBaseLength();
    //每次转向的角度
    void SetAngle(double angle);
    double GetAngle();
    //计算包围盒
    void CalculateBoundingBox();
    //设置初始枝干的角度
    void SetInitDirectionAngle(double angle);
    //对任意符号串的绘制
    void Draw(RECT rect, HDC hdc, const std::string& curveStr);
```

```
        //绘制对外接口，只绘制 m_n 迭代生成的符号串
        void Draw(RECT rect, HDC hdc);
private:
        //从 1,...,m_n 生成所有符号字符串
        std::vector<std::string> m_curveStrings;
        //每次枝干转向的角度
        double m_angle;
        //初始枝干的角度
        double m_initDirAngle;
        //每次枝干生长的长度
        double m_baseLength;
        //迭代次数
        int m_n;
        //包围盒
        RECT m_BoundingBox;
};
```

FractalTree 类中符号字符串生成函数定义如下：


```
/*
变量 : X F
常量  : +-[]
原子符号   : X
生成规则   : (X → F[-X][X]F[-X] + FX), (F → FF)
这里，F 表示向前绘制一段距离, -表示左转 25°，+表示右转 25°。X 变量和绘制内容无关，
*/
void FractalTree::GenerateString()
{
        int i, j;
        std::string lastStr, str;
        m_curveStrings.clear();
        lastStr = str = "X";
        m_curveStrings.push_back(str);
        for (i = 2; i <= m_n; ++i)
        {
                str.clear();
                for (j = 0; j < (int)lastStr.size(); ++j)
                {
                        if (lastStr[j] == 'X')
                                str += "F[-X][X]F[-X]+FX";
                        else if (lastStr[j] == 'F')
                                str += "FF";
                        else
                                str += lastStr[j];
```

```
            }
            m_curveStrings.push_back(str);
            lastStr = str;
        }
    }
```

GenerateString 函数中，m_curveStrings 是一个字符串 vector 容器对象，它存储了从 1 迭代到 n 的所有符号字符串。当然在绘制时只需要绘制最后一个字符串，代码如下：

```
//绘制的对外接口，只绘制 m_n 迭代生成的符号串
void FractalTree::Draw(RECT rect, HDC hdc)
{
    if (m_curveStrings.empty())
        return;
    Draw(rect, hdc, m_curveStrings.back());
}
```

符号串绘制函数 Draw 定义如下。其中 std::vector<ParaInfor> paraStack 就是一个参数堆栈，当然也可以用 std::stack<ParaInfor>定义该堆栈。需要注意的是调用 Draw 之前，一定要调用 CalculateBoundingBox 函数计算符号串分形树的包围盒，然后通过坐标映射的方法将包围盒下的坐标映射到实际绘制窗口下的坐标。坐标映射的方法非常简单：缩放+平移，代码如下：

```
void FractalTree::Draw(RECT rect, HDC hdc, const std::string& curveStr)
{
    int i, yBottom;
    //绘制树干需要维护的堆栈，堆栈的内容就是 ParaInfor，即二维点坐标和二维方向向量
    std::vector<ParaInfor> paraStack;
    //当前二维点坐标和二维方向向量
    ParaInfor curData;
    double dirX, dirY, offsetX, offsetY, scaleX, scaleY, scale=1.0;
    //坐标映射需要用到的缩放比例 scale，以及平移参数 offsetX 和 offsetY
    scaleX  =
    (rect.right-rect.left+0.0)/(m_BoundingBox.right-m_BoundingBox.left);
    scaleY  =
    (rect.bottom-rect.top+0.0)/(m_BoundingBox.bottom-m_BoundingBox.top);
    scale = std::min(scaleX, scaleY)*0.8;
    //计算平移参数
    offsetX =
    (rect.left+rect.right)*0.5-(m_BoundingBox.right+m_BoundingBox.left)*scale*0.5;
    offsetY =
    (rect.bottom+rect.top)*0.5-(m_BoundingBox.bottom+m_BoundingBox.top)*scale*0.5;
    //从底部绘制，由于 GDI 坐标系的 Y 轴是反向的，所以这里设置底部为 rect.bottom
    yBottom = rect.bottom;
    //设置初始二维点坐标和二维方向向量
    curData.x = 0;
    curData.y = 0;
```

```
//设置初始方向（树主干方向）
curData.dirX = cos(m_initDirAngle*PI/180);
curData.dirY = sin(m_initDirAngle*PI / 180);
//准备旋转矩阵数据
double cosA, sinA, cosB, sinB;
cosA = cos(PI / 180 * m_angle);
sinA = sin(PI / 180 * m_angle);
cosB = cos(-PI / 180 * m_angle);
sinB = sin(-PI / 180 * m_angle);
//开始解析符号串，规则如下
/*
F 表示向前绘制一段距离,-表示左转 25°，+表示右转 25°,X 变量和绘制内容无关，[压栈，]退栈
*/
for (i = 0; i < (int)curveStr.size(); ++i)
{
    if (curveStr[i] == 'F')//向前绘制一段距离
    {
        //将 curData 中的坐标映射到 rect 窗口坐标系下：坐标映射 = 缩放+平移
        //由于 GDI 坐标系下 Y 轴是反向的，所以需要用 yBottom 减去映射后的 y 坐标
        ::MoveToEx(hdc, (int)(curData.x*scale+offsetX),
                    (int)(yBottom-(curData.y*scale+offsetY)), NULL);
        ::LineTo(hdc, (int)((curData.x + curData.dirX*m_baseLength)*scale+offsetX),
        (int)(yBottom - ((curData.y + curData.dirY*m_baseLength)*scale+offsetY)));
        //并把当前位置挪动到最新位置
        curData.x += curData.dirX*m_baseLength;
        curData.y += curData.dirY*m_baseLength;
    }
    else if (curveStr[i] == '-') //-表示左转 25°
    {
        //利用旋转公式,对当前方向进行转向
        /*
        | cos(a)   -sin(a)|     |x|      |x'|
        |                 | * | |  =  |  |
        | sin(a)   cos(a) |     |y|      |y'|
        */
        dirX = cosA*curData.dirX - sinA*curData.dirY;
        dirY = sinA*curData.dirX + cosA*curData.dirY;
        curData.dirX = dirX;
        curData.dirY = dirY;
    }
    else if (curveStr[i] == '+') //-表示右转 25°
    {
        dirX = cosB*curData.dirX - sinB*curData.dirY;
```

```
                dirY = sinB*curData.dirX + cosB*curData.dirY;
                curData.dirX = dirX;
                curData.dirY = dirY;
            }
            else if (curveStr[i] == '[') //将当前位置和方向压入栈
            {
                paraStack.push_back(curData);
            }
            else if (curveStr[i] == ']') //将栈顶元素退栈，重置当前位置和方向
            {
                assert(!paraStack.empty());
                curData = paraStack.back();
                paraStack.pop_back();
            }
        }
    }
```

CalculateBoundingBox 函数计算符号串分形树的包围盒，它的基本代码和 Draw 非常类似，只是去掉了绘制部分的代码，代码如下：

```
void FractalTree::CalculateBoundingBox()
{
    m_BoundingBox.left = 10000;
    m_BoundingBox.right = -10000;
    m_BoundingBox.top = 10000;
    m_BoundingBox.bottom = -10000;
    if(m_curveStrings.empty())
        return;
    int i;
    //绘制树干需要维护的堆栈，堆栈的内容就是ParaInfor，即二维点坐标和二维方向向量
    std::vector<ParaInfor> paraStack;
    std::string curveStr = m_curveStrings.back();
    //当前二维点坐标和二维方向向量
    ParaInfor curData;
    double dirX, dirY;

    //设置初始二维点坐标和二维方向向量
    curData.x = 0;
    curData.y = 0;
    //设置初始方向
    curData.dirX = cos(m_initDirAngle*PI/180);
    curData.dirY = sin(m_initDirAngle*PI / 180);

    //开始解析符号串，规则如下
    /*
```

```
F 表示向前绘制一段距离，-表示左转 25°，+表示右转 25°，X 变量和绘制内容无关，[压栈，]退栈
*/
for (i = 0; i < (int)curveStr.size(); ++i)
{
    if (curveStr[i] == 'F')//向前绘制一段距离
    {
        if(curData.x<m_BoundingBox.left)
            m_BoundingBox.left = curData.x;
        if(curData.x>m_BoundingBox.right)
            m_BoundingBox.right = curData.x;
        if(curData.y<m_BoundingBox.top)
            m_BoundingBox.top = curData.y;
        if(curData.y>m_BoundingBox.bottom)
            m_BoundingBox.bottom = curData.y;

        if(curData.x +
     curData.dirX*m_baseLength<m_BoundingBox.left)
            m_BoundingBox.left  =  curData.x + curData.dirX*m_baseLength;
        if(curData.x  +
     curData.dirX*m_baseLength>m_BoundingBox.right)
            m_BoundingBox.right = curData.x +
     curData.dirX*m_baseLength;
        if(curData.y + curData.dirY*m_baseLength<m_BoundingBox.top)
            m_BoundingBox.top = curData.y + curData.dirY*m_baseLength;
        if(curData.y +
     curData.dirY*m_baseLength>m_BoundingBox.bottom)
            m_BoundingBox.bottom = curData.y +
      curData.dirY*m_baseLength;

        //并把当前位置挪动到最新位置
        curData.x += curData.dirX*m_baseLength;
        curData.y += curData.dirY*m_baseLength;
    }
    else if (curveStr[i] == '-') //-表示左转 25°
    {
        //利用旋转公式,对当前方向进行转向
        /*
        | cos(a)   -sin(a)|     |x|     |x'|
        |                       |*| | =|  |
        | sin(a)   cos(a) |     |y|     |y'|
        */
        dirX = cos(PI / 180 * m_angle)*curData.dirX - sin(PI / 180 * m_angle)*curData.dirY;
        dirY = sin(PI / 180 * m_angle)*curData.dirX + cos(PI / 180 * m_angle)*curData.dirY;
        curData.dirX = dirX;
```

```
                curData.dirY = dirY;
            }
            else if (curveStr[i] == '+') //-表示右转 25°
            {
                dirX = cos(-PI / 180 * m_angle)*curData.dirX - sin(-PI / 180 * m_angle)*curData.dirY;
                dirY = sin(-PI / 180 * m_angle)*curData.dirX + cos(-PI / 180 * m_angle)*curData.dirY;
                curData.dirX = dirX;
                curData.dirY = dirY;
            }
            else if (curveStr[i] == '[') //将当前位置和方向压入栈
            {
                paraStack.push_back(curData);
            }
            else if (curveStr[i] == ']') //将栈顶元素退栈，重置当前位置和方向
            {
                assert(!paraStack.empty());
                curData = paraStack.back();
                paraStack.pop_back();
            }
        }
    }
```


最后设计 main.cpp 文件中的界面和框架代码。首先定义三个全局变量：分形树对象、是否使用双缓冲的标志、当前迭代次数标志。其中，双缓冲是为了加快 GDI 的显示速度，因为随着迭代次数增大需要绘制的树枝分叉数越来越多，双缓冲可以加速树的绘制过程，代码如下：

```
//分形树
FractalTree g_Tree;
//初始不使用双缓冲
bool g_bDoubleBuffer=false;
//初始迭代次数
int g_N = 5;
```

定义菜单项。菜单中提供迭代次数和双缓冲的设置项，代码如下：

```
//声明菜单项 ID
#define IDM_N    1
#define IDM_N_1    2
#define IDM_N_MAX    14
#define IDM_DOUBLE_BUFFER 15

//菜单栏—菜单—菜单项
void CreateMainMenu(HWND hWnd)
{
    HMENU hMenuBar;                    // 菜单栏
```

```
    HMENU hMenuParameter;          // 参数菜单
    HMENU hMenuParameterN;         // 迭代次数菜单
    int i;
    wchar_t buffer[16];
    //创建菜单栏
    hMenuBar = CreateMenu();

    //创建参数菜单
    hMenuParameter = CreateMenu();
    //创建迭代次数弹出菜单
    hMenuParameterN = CreatePopupMenu();
    //在迭代次数菜单下，添加菜单项 IDM_N_1 至 IDM_N_MAX
    for(i=IDM_N_1; i<=IDM_N_MAX; ++i)
    {
        swprintf(buffer, L"&%d", i-1);
        AppendMenuW(hMenuParameterN, MF_UNCHECKED, i, buffer);
    }
    //将迭代次数菜单添加到参数菜单下
    AppendMenuW(hMenuParameter, MF_POPUP, (UINT_PTR)hMenuParameterN, L"&Parameter");

    //在参数菜单下，添加菜单项双缓冲
    AppendMenuW(hMenuParameter, MF_UNCHECKED, IDM_DOUBLE_BUFFER, L"&Double Buffer");

    //将参数菜单添加到主菜单栏
    AppendMenuW(hMenuBar, MF_POPUP, (UINT_PTR)hMenuParameter, L"&Parameter");

    //将主菜单栏设置到当前窗口
    SetMenu(hWnd, hMenuBar);
    return;
}
```


主窗口消息回调函数定义如下：

```
LRESULT CALLBACK WindowProcedure (HWND hwnd, UINT message, WPARAM wParam, LPARAM
lParam)
{
    switch (message)                    /* handle the messages */
    {
    case WM_DESTROY:
        PostQuitMessage (0);       /* send a WM_QUIT to the message queue */
        break;
    case WM_CREATE:   //窗口创建时发送的消息，在该消息中我们初始化分形树参数
        {
            CreateMainMenu(hwnd);
            HMENU hMenu = GetMenu(hwnd);
```

```
                CheckMenuItem(hMenu, IDM_N_1+g_N-1, MF_CHECKED);
                g_Tree.SetN(g_N);                           //设置迭代次数
                g_Tree.SetBaseLength(10);                   //设置生长的长度
                g_Tree.SetInitDirectionAngle(70);           //设置初始树干方向角度
                g_Tree.GenerateString();                    //生成符号串
                g_Tree.CalculateBoundingBox();              //计算包围盒
            }
            break;
        case WM_COMMAND:
            {
                HMENU hMenu = GetMenu(hwnd);
                if(wParam==IDM_DOUBLE_BUFFER)
                {
                    g_bDoubleBuffer = !g_bDoubleBuffer;
                    CheckMenuItem(hMenu, IDM_DOUBLE_BUFFER,g_bDoubleBuffer?MF_CHECKED:
MF_ UNCHECKED);
                }
                else if(wParam>=IDM_N_1 && wParam<=IDM_N_MAX)
                {
                    CheckMenuItem(hMenu, IDM_N_1+g_N-1, MF_UNCHECKED);
                    g_N = wParam-1;
                    CheckMenuItem(hMenu, IDM_N_1+g_N-1, MF_CHECKED);
                    g_Tree.SetN(g_N);                       //设置迭代次数
                    g_Tree.GenerateString();                //生成符号串
                    g_Tree.CalculateBoundingBox();          //计算包围盒
                }
                InvalidateRect(hwnd, NULL, 1);
            }
            break;
        case WM_SIZE:
            InvalidateRect(hwnd, NULL, 1);
            break;
        case WM_PAINT:
            {
                PAINTSTRUCT ps;
                RECT rect;
                HDC hdc,hMemDC;
                HBITMAP hBmp;

                hMemDC = hdc = BeginPaint(hwnd, &ps);
                GetClientRect(hwnd, &rect);

                if(g_bDoubleBuffer)
                {
```

```
                    hMemDC = CreateCompatibleDC(hdc);
                    hBmp = CreateCompatibleBitmap(hdc, rect.right, rect.bottom);
                    SelectObject(hMemDC,hBmp);
                    FillRect(hMemDC,&rect, GetSysColorBrush(COLOR_3DFACE));
                }
                g_Tree.Draw(rect, hMemDC); //绘制分形树

                if(g_bDoubleBuffer)
                {
BitBlt(hdc,0,0,rect.right,rect.bottom,hMemDC,0,0,SRCCOPY);
                    DeleteObject(hBmp);
                    DeleteDC(hMemDC);
                }
                EndPaint(hwnd, &ps);
            }
            break;
        default:                        /* for messages that we don't deal with */
            return DefWindowProc (hwnd, message, wParam, lParam);
    }
    return 0;
}
```


当迭代次数 *n* 设置为 13 时，分形植物如图 8-11 所示。

图 8-11　分形植物（*n*=13）

8.4 迷宫问题

迷宫问题包含两个子问题：迷宫生成算法与迷宫行走算法。数据结构或算法书中通常只讨论如何走出迷宫，但迷宫生成实际上也是非常有趣的问题。

1. 迷宫生成

迷宫生成有若干种随机生成算法，这里只讨论随机深度优先搜索（Random DFS）与随机宽度优先搜索（Random BFS）两种算法。

Random DFS 算法采用堆栈或递归方式生成迷宫。随机性主要体现在将当前位置的左、上、右、下四个方向组合成一个数组，然后进行随机打乱顺序，随后进行行走测试。特别强调的是，迷宫生成时总是走两步。例如，如果往右走，那么需要测试紧邻的右边两个位置是否都没有走过，只有都没有访问才会往右走两步。基于上述事实，迷宫生成的 Random DFS 算法流程如下。

（1）初始化地图上所有格子都为墙（二维数组元素都为 0）。

（2）标记起点所在的格子为通路（设置起点位置数组元素值为 1），并将当前位置设置为起点。

（3）开始如下循环。

① 将上、下、左、右四个方向打乱，然后存到含有 4 个元素的一维方向数组中。

② 设置一个走动成功标志，当前标志值为 0，表示还未成功。

③ 循环这个一维方向数组。测试从当前位置往当前方向行走两个格子，如果这两个格子都是墙，那么将这两个格子都设置为通路；同时将当前位置压到堆栈中；将当前方向走两格的那个位置设置为当前位置；最后设置走动成功标志为 1，退出当前一维方向数组的循环。

④ 如果走动成功标志为 0，表示行走失败，则测试堆栈是否为空，如果堆栈为空，则退出整个循环。否则从堆栈中弹出一个元素作为当前位置。

基于上述算法，迷宫生成的 Random DFS 代码如下：

```
void Maze::GenerateMaze_DepthFirstSearch()
{
    std::vector<std::pair<int, int> > visitStacks;

    int nextM, nextN, dir, mm, nn, mm2, nn2;
    int i, j;
    static int dirsX[4]= {2,0,-2,0};
    static int dirsY[4]= {0,2,0,-2};
    static int dirsX2[4]= {1,0,-1,0};
    static int dirsY2[4]= {0,1,0,-1};
    int dirs[4]={0,1,2,3};

    //清除原有地图
    ClearMapContent();
```

```
//设置 4 条虚拟边界
for(i=0; i<m_n; ++i)
{
    m_map[0][i] = 1;
    m_map[m_m-1][i] = 1;
}
for(i=0; i<m_m; ++i)
{
    m_map[i][0] = 1;
    m_map[i][m_n-1] = 1;
}
//将起点设置为通路
m_map[m_StartM][m_StartN] = 1;
nextM = m_StartM;
nextN = m_StartN;
while(true)
{
    //生成随机方向数组
    for(i=0; i<4; ++i)
    {
        j=rand()%4;
        std::swap(dirs[i], dirs[j]);
    }
    bool bSucc=false;
    for(j=0; j<4; ++j)
    {
        mm = nextM+dirsX[dirs[j]];
        nn = nextN+dirsY[dirs[j]];
        mm2 = nextM+dirsX2[dirs[j]];
        nn2 = nextN+dirsY2[dirs[j]];
        //探测是否越界
        if(mm>=0 && mm<m_m && nn>=0 && nn<m_n)
        {
            //如果两个格子都是墙
            if(!m_map[mm][nn] && !m_map[mm2][nn2])
            {
                //将两个格子都设置为通路
                m_map[mm][nn] = 1;
                m_map[mm2][nn2] = 1;
                //把当前位置压到堆栈中
                visitStacks.push_back(std::make_pair(nextM, nextN));
                //更新当前位置
```

```
                    nextM=mm;
                    nextN=nn;
                    //设置访问成功标志位
                    bSucc = true;
                    break;
                }
            }
        }
        //如果4个方向的探测都失败
        if(!bSucc)
        {
            //如果堆栈不为空
            if(!visitStacks.empty())
            {
                //将栈顶元素设置成当前位置
                nextM = visitStacks.back().first;
                nextN = visitStacks.back().second;
                //探出栈顶元素
                visitStacks.pop_back();
            }
            else
            {
                //堆栈为空，退出整个循环
                break;
            }
        }
    }
    //如果出口并不在通路上，则确保出口和通路上有一条路径
    if(!m_map[m_EndM][m_EndN])
    {
        m_map[m_EndM][m_EndN] = 2;

        bool bInPath = false;
        std::vector<std::pair<int, int> >patchPath; //exit related extra path to main path
        nextM = m_EndM;
        nextN = m_EndN;
        patchPath.push_back(std::make_pair(nextM, nextN));
        while(!bInPath)
        {
            for(i=0; i<4; ++i)
            {
                mm = nextM+dirsX2[i];
                nn = nextN+dirsY2[i];
```

```
                    if(mm>0 && mm<m_m-1 && nn>0 && nn<m_n-1 && m_map[mm][nn]==1)
                    {
                        bInPath = true;
                        break;
                    }
                }
                if(!bInPath)
                {
                    for(i=0; i<4; ++i)
                    {
                        mm = nextM+dirsX2[i];
                        nn = nextN+dirsY2[i];
                        if(mm>0 && mm<m_m-1 && nn>0 && nn<m_n-1)
                        {
                            nextM = mm;
                            nextN = nn;
                            m_map[nextM][nextN] = 2;
                            patchPath.push_back(std::make_pair(nextM, nextN));
                            break;
                        }
                    }
                }
            }
            for(i=0; i<patchPath.size(); ++i)
                m_map[patchPath[i].first][patchPath[i].second] = 1;
        }

        for(i=0; i<m_n; ++i)
        {
            m_map[0][i] = 0;
            m_map[m_m-1][i] = 0;
        }
        for(i=0; i<m_m; ++i)
        {
            m_map[i][0] = 0;
            m_map[i][m_n-1] = 0;
        }
    }
```

Random DFS 算法本身并不指定入口和出口位置，所以在上面代码中，最后还需要从终点位置随机反向访问现有的通路。更确切地说，就是要把终点和现有通路上随机的一个位置打通。Random DFS 迷宫生成效果如图 8-12 所示。

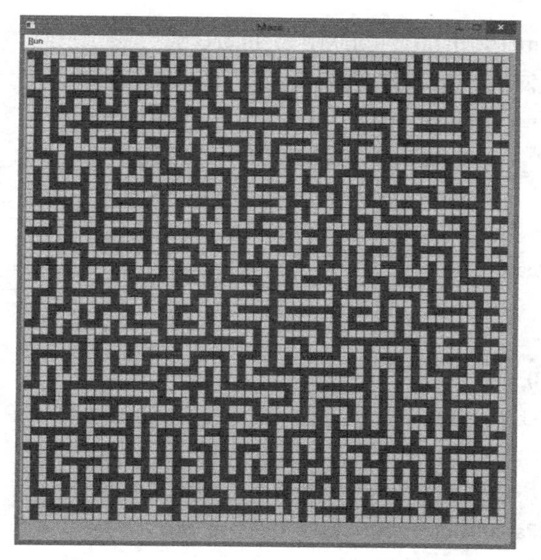

图 8-12 Random DFS 迷宫生成效果

迷宫生成的 Random BFS 算法和 Random DFS 算法非常类似，只需要将堆栈数据结构改成队列即可，这里不再赘述。Random BFS 迷宫生成效果如图 8-13 所示。

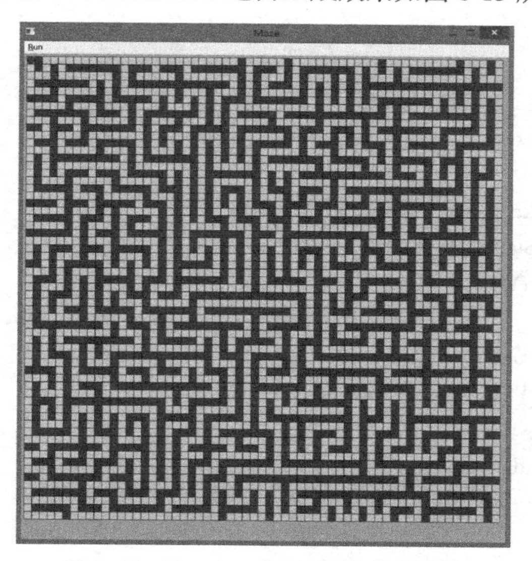

图 8-13 Random BFS 迷宫生成效果

在实际测试中可以观察 Random BFS 生成的迷宫要比 Random DFS 迷宫复杂，因此推荐使用 Random BFS 算法生成迷宫。

2. 迷宫行走

在第 4 章中，已经给出基于 BFS 的走迷宫最短路径算法。算法要点是要多分配一个 M*N 的方向数组，用来记录每个格子是从哪个方向走过来的。当 BFS 遍历整个迷宫后，该方向数组会填满方向信息。此时只需要从该方向数组的终点反向往回推到起点，从而得到一系列格子坐标，最后输出这些坐标即可。代码如下：

```
/*入口(m_StartM,m_StartN)，出口(m_EndM，m_EndN)
m_map[i][j]==0 表示该网格不通，否则表示通
*/
enum GRID_DIR { DIR_NONE=0, DIR_DOWN, DIR_LEFT, DIR_UP, DIR_RIGHT};
void Maze::Solve()
{
    m_SearchedPath.clear();
    if(!m_map)
        return;
    static int dirsX[]={0, 1, 0, -1};
    static int dirsY[]={-1, 0, 1, 0};
    int i, m, n, nextM, nextN;
    //二维方向数组，用来记录每个格子是从哪个方向走过来的
    GRID_DIR **gridParentDirs = new GRID_DIR*[m_m];
    for(i=0;i<m_m; ++i)
    {
        gridParentDirs[i] = new GRID_DIR[m_n];
        memset(gridParentDirs[i], 0, sizeof(GRID_DIR)*m_n);
    }
    //BFS 队列
    std::queue<std::pair<int, int> > seeds;
    std::pair<int, int> seed;
    bool bFindIt=false;
    seeds.push(std::make_pair(m_StartM, m_StartN));
    gridParentDirs[m_StartM][m_StartN] = DIR_DOWN;
    while(!seeds.empty())
    {
        seed = seeds.front();
        seeds.pop();
        m = seed.first;
        n = seed.second;
        for(i=0; i<4; ++i)
        {
            nextM = m+dirsX[i];
            nextN = n+dirsY[i];
            if(m_map[nextM][nextN] && gridParentDirs[nextM][nextN]==DIR_NONE)
            {
                seeds.push(std::make_pair(nextM, nextN));
                gridParentDirs[nextM][nextN] = GRID_DIR(i+1);
                if(nextM==m_EndM && nextN==m_EndN)
                {
                    bFindIt=true;
                    break;
                }
```

```
                }
            }
            if(bFindIt)
                break;
        }
        //反向找出路径
        m_SearchedPath.push_back(std::make_pair(m_EndM, m_EndN));
        seed = std::make_pair(m_EndM, m_EndN);
        while(true)
        {
            if(seed.first= =m_StartM && seed.second==m_StartN)
                break;
            i = (int)gridParentDirs[seed.first][seed.second]-1;
            seed.first = seed.first-dirsX[i];
            seed.second = seed.second-dirsY[i];
            m_SearchedPath.push_back(seed);
        }

        for(i=0; i<m_m; ++i)
            delete[] gridParentDirs[i];
        delete[] gridParentDirs;
    }
```

BFS 路径求解结果如图 8-14 所示。

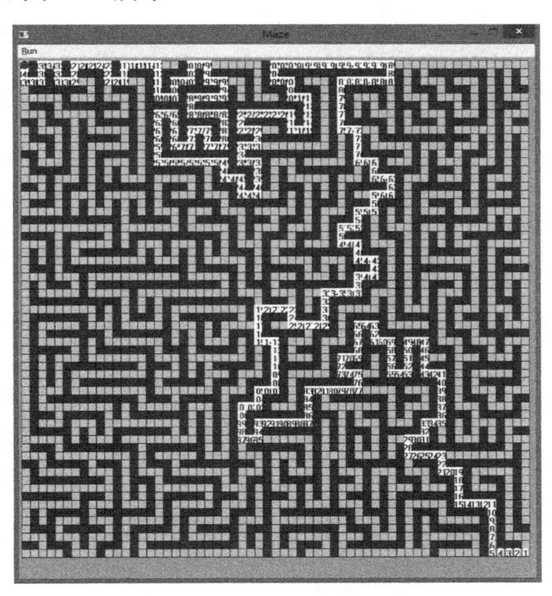

图 8-14 BFS 路径求解结果

详细算法请读者参考代码。

参考文献

[1] 刘汝佳. 算法竞赛入门经典（第 2 版）[M]. 北京：清华大学出版社，2014.

[2] PTA（https://pintia.cn/.）

[3] PAT（https://www.patest.cn/contests.）

[4] 陈越，何钦铭，等. 数据结构（第 2 版）[M]. 北京：高等教育出版社，2016.

[5] C++ Manual（http://www.cplusplus.com/.）

[6] 张新华. 算法竞赛宝典· 第二部：基础算法艺术[M]. 北京：清华大学出版社，2016.

[7] C++ Reference（http://en.cppreference.com/w/.）

[8] Nicolai M. Josuttis. C++标准库（第 2 版）[M]. 北京：电子工业出版社，2015.

[9] L-System（https://en.wikipedia.org/wiki/L-system.）

[10] Win32 GDI（https://msdn.microsoft.com.）

反侵权盗版声明